Sprayed Concrete Linings in Soft Ground

Sprayed Concrete Linings in Soft Ground

A best practice design guide

The British Tunnelling Society

Published by Emerald Publishing Limited, Floor 5, Northspring, 21–23 Wellington Street, Leeds LS1 4DL.

ICE Publishing is an imprint of Emerald Publishing Limited

Other ICE Publishing titles:
Specification for Tunnelling, Fourth edition
The British Tunnelling Society. ISBN 978-0-7277-6643-4
Rock Engineering
Håkan Stille and Arild Palmström. ISBN 978-0-7277-4083-0
Crossrail Project: Infrastructure design and construction
Edited by Mike Black, Christian Dodge and Ursula Lawrence.
ISBN: 978-0-7277-6078-4

A catalogue record for this book is available from the British Library

ISBN 978-1-8360-8693-2

© British Tunnelling Society 2025. Published by Emerald
Publishing Limited.

Cover photo: Bank Station Capacity Upgrade platform junctions.
Photo courtesy of Dragados and London Underground.
Commissioning Editor: Michael Fenton
Content Development Editor: Ryan Molyneux
Books Production Lead: Benn Linfield

Typeset by KnowledgeWorks Global Limited
Index created by David Gaskell

Contents

Foreword

The idea for this guide was sparked over 13 years ago, when Ross Dimmock (Normet) met with Dr Keith Bowers (Cowi, formally Transport for London) and Bob Ibell (then Chair of the British Tunnelling Society) for a quick beverage after a meeting. The Crossrail Project was in full flow, and knowledge of sprayed concrete linings (SCL) in soft ground was advancing in leaps and bounds. It was decided that this knowledge should be captured in a guide to all things SCL: covering design, construction, materials, equipment, safety and much more. Support was gathered at a subsequent meeting of the British Tunnelling Society Committee and a technical working group established. As everyone was so busy it was feared that it might take a couple of years to complete, but enthusiasm was high.

The task was split into different chapters, with a lead assigned to each. It was at this point that I signed up to contribute, requesting to join the design team, led by Brian Lyons. The team quickly established a list of topics to be included within the design chapter, complete with bullet point lists to be expanded under each heading. These were submitted for review, with the hope that the different chapters would be aligned to avoid repetition and overlap.

Time passed. Everyone was indeed very busy, and the other chapters of the book appeared to be lagging behind. It was also noted that there were quite a few topics under the design heading, and that when combined with the other chapters, it might be a very long book indeed. The *Sprayed Concrete Linings in Soft Ground: a best practice design guide* was born. Much was completed by 2019, but sadly the pandemic diverted resources from the final review and publication. That the final reviews were completed during my tenure as chair of the BTS Technical Sub-committee feels fitting, and I am very pleased to have ushered it over the line and share it with you all.

This guide gathers knowledge from the Crossrail Project, where SCL was employed to create the five awe-inspiring central stations (Bond Street, Tottenham Court Road, Farringdon, Liverpool Street and Whitehchapel) as well as crossovers and cross passages under the streets of central London. This was supplemented by concurrent and subsequent projects, including upgrades to Transport for London's stations (including the Bond Street Station Upgrade, Tottenham Court Road Station Upgrade, Victoria Station Upgrade, Bank Station Capacity Upgrade and the Northern Line Extension), the Thames Tideway Tunnel and other utility upgrades, right up to the ongoing High Speed 2 works north of Euston Station.

Sprayed concrete linings were used in all these projects due to their ability to adapt to complex requirements. Working among the maze of existing underground infrastructure of London requires tunnel profiles to change and turn in ways that are impractical using tunnel boring machines. The tunnels are equally too large to complete using hand-mining techniques alone. How SCL is used here in London does differ from its roots in hard rock tunnels, and this guide is specifically looking to capture the considerations of SCL tunnelling in soft ground. Where a topic warrants greater depth than we have space for, we have signposted useful references for further reading. We hope that it serves as a useful introduction to engineers, clients and any other interested parties and helps give confidence in the use of this fantastic material for future tunnelling works.

Bethan Haig
BTS Technical Subcommittee Chair 2023–2025

Acknowledgments

This best practice design guide was prepared by the SCL Tunnel Design in Soft Ground working group of the British Tunnelling Society Technical subcommittee.

Members of the committee:
Brian Lyons, Dr Sauer & Partners
Christoph Eberle, Mott MacDonald
Nick Tucker, Mott MacDonald
Dr Benoit Jones, BEMO Tunnelling
Bethan Haig, Dr Sauer & Partners
Dr Ali Nasekhian, Dr Sauer & Partners
Adrian St John, Kier Group PLC

Peer reviewers:
Mike King, MK Tunnelling Limited
Neil Moss, Gall Zeidler
Frank Mimnagh, Cowi

emerald
PUBLISHING

ice

British Tunnelling Society
ISBN 978-1-83608-693-2
https://doi.org/10.1108/978-1-83608-690-120251001

Chapter 1
Introduction

The objectives of this best practice design guideline are as follows:

- Provide the designer with a list of current industry standards, guidelines and specifications relevant to sprayed concrete lining (SCL) design and construction.
- Provide guidance in the selection of SCL configuration (e.g. double shell, composite, single shell).
- Provide guidance on spaceproofing and tolerances for SCL construction.
- Provide guidance on ground investigation and the selection of ground and water parameters.
- Provide typical details for excavation sequence and face division, together with a narrative describing their relative benefits.
- Provide guidance for the choice of material parameters and methodologies for SCL design, including designing for fire.
- Provide guidance on ground movements and instrumentation and monitoring for SCL.
- Provide guidance on sustainable design approaches for SCL.
- Provide guidance on the design specification and assurance processes.

Health and safety is integral to every part of the design process.

This guideline has been developed for the benefit of designers, project managers, contractors and clients involved in the design and construction of SCL infrastructure.

It is targeted at the design and construction of SCL in soft ground, London Clay in particular. It is not generally applicable to hard rock conditions. It has been influenced by recent developments within the industry, most notably the successful delivery of the Elizabeth Line (Crossrail) in London and capacity upgrades for Transport for London, which involved extensive use of sprayed concrete lined tunnels.

emerald PUBLISHING ice

British Tunnelling Society
ISBN 978-1-83608-693-2
https://doi.org/10.1108/978-1-83608-690-120251002

Chapter 2
Standards and codes of practice

A number of publications have been developed over the years in the UK, Europe and the USA that have been adopted for the use in SCL design.

2.1.   Limitations in current standards

There are few standards and codes specifically developed for the use of SCL. The following areas are particularly under-developed:

- Lack of SCL-specific design standards or codes
- Ground structure interaction using numerical modelling (clarity on use of load and material factors)
- Residual flexural tensile capacity and ductility of sprayed fibre reinforced concrete
- Tensile capacity of plain and fibre reinforced concrete
- Composite linings (SCL/membrane/SCL)
- Analysis of radial joints in SCL
- Permanent/temporary nature of SCL
- Advance rates in relation to strength gain.

2.2.   General guidelines

- *fib Model Code for Concrete Structures 2010* (*fib*, 2013).
- *Sprayed Concrete Linings (NATM) for Tunnels in Soft Ground.* Institution of Civil Engineers design and practice guides (ICE, Thomas Telford, 1996).
- *Tunnel Lining Design Guide* (British Tunnelling Society (BTS), 2004).
- *TR63 Guidance for the Design of Steel-Fibre-Reinforced Concrete* (Concrete Society, 2007).
- *Guideline Shotcrete* (Austrian Concrete Society, 2013). This publication applies to the production of structural components made of plain and reinforced concrete as well as the close-textured reinforced concrete placed by the method of spraying. The document covers the testing and design of sprayed concrete mixes, and does so by investigating the requirements placed on the finished material.
- *Safety of New Austrian Tunnelling Method (NATM) Tunnels* (Health and Safety Executive (HSE), 1996a). In October 1994, tunnels under construction beneath Heathrow Airport collapsed. The HSE considered whether there were any broader health and safety implications concerning both the construction of NATM tunnels in the UK and the safety of the finished tunnel in comparison with traditional methods.
- *Post Construction Audit of Sprayed Concrete Tunnel Linings* (HSE, 1996b). This report researches the practice in the auditing of the finished sprayed concrete tunnel lining structure.

- *Low Carbon Routemap* (Low Carbon Concrete Group/The Green Construction Group, ICE, 2022).
- *Design Guidance for Spray Applied Waterproofing Membranes* (ITAtech, 2013).
- *Guidance for Precast Fibre Reinforced Concrete Segments, Volume 1: Design Aspects* (ITAtech, 2016).
- *Low Carbon Concrete Linings* (ITAtech, 2024).
- *Permanent Sprayed Concrete Linings* (Working Group No. 12, ITA, 2020).
- *How to Calculate Embodied Carbon* (IStructE, 2022).
- ACI 506R-16: Guide to Shotcrete (ACI, 2016).
- ACI 544.4-18: Guide to Design with Fiber-Reinforced Concrete (ACI, 2018).
- ACI 544.5R-10: Report on the Physical Properties and Durability of Fiber-Reinforced Concrete (ACI, 2010).
- *Monitoring Underground Construction: A Best Practice Guide* (BTS, 2011).
- A Code of Practice for Risk Management of Tunnel Works (International Tunnelling and Underground Space Association (ITA-AITES) International Association of Engineering Insurers (IMA), 2023)

2.3. Eurocodes and British standards

- BS EN 1990: Eurocode 0. Basis of structural design.
- BS EN 1991-1: Eurocode 1. Actions on structures.
- BS EN 1992-1: Eurocode 2. Design of concrete structures.
- BS EN 1997-1: Eurocode 7. Geotechnical design. General rules.
- BS EN 1997-2: Eurocode 7. Geotechnical design. Ground investigation and testing.
- BS EN 14487-1:2022: Sprayed concrete. Part 1: Definitions, specifications and conformity.
- BS EN 14487-2:2006: Sprayed concrete. Part 2: Execution.
- BS EN 14488-1:2005: Testing sprayed concrete. Sampling fresh and hardened concrete.
- BS EN 14488-2:2006: Testing sprayed concrete. Compressive strength of young sprayed concrete.
- BS EN 14889-1:2006: Fibres for concrete. Steel fibres. Definitions, specifications and conformity.
- BS EN 14889-2:2006: Fibres for concrete. Polymer fibres. Definitions, specifications and conformity.
- BS EN 14651:2005: Test method for metallic fibre concrete. Measuring the flexural tensile strength (limit of proportionality (LOP), residual).
- BS EN ISO 14688-1: Geotechnical investigation and testing. Identification and classification of soil. Identification and description.
- BS EN ISO 14688-2: Geotechnical investigation and testing. Identification and classification of soil. Principles for a classification.
- BS EN 206-1: Concrete. Specification, performance, production and conformity.
- BS 6164:2019: Health and safety in tunnelling in the construction industry – Code of practice.
- BS 8500-1: Concrete. Complementary British Standard to BS EN 206. Method of specifying and guidance for the specifier.
- BS 8500-2: Concrete. Complementary British Standard to BS EN 206. Specification for constituent materials and concrete.
- ISO 44001: Collaborative business relationship management systems. Requirements and framework.

- ISO 834-1: Fire resistance tests.
- The Construction (Design and Management) Regulations 2015 (CDM).

2.4. Flexural strength guidelines

- RILEM TC 162-TDF (2003) Test and design methods for steel fibre reinforced concrete. *Materials and Structures* **36**: 560–567.
- *fib Model Code for Concrete Structures*. Bulletin 65, vol. 1, Section 5.6: Fibres and fibre reinforced concrete (*fib*, 2010).

2.5. Specifications

- *European Specification for Sprayed Concrete* (EFNARC, 1996). This specification treats sprayed concrete as an entity and makes no reference to fields of application, such as tunnelling, which is the case in many other publications.
- *European Specification for Sprayed Concrete: Guidelines for Specifiers & Contractors* (EFNARC, 1999). This publication is to be read in conjunction with the EFNARC *Specification for Sprayed Concrete*. These guidelines refer to the specification and all references relate to the clause numbers in the specification. The guidelines contain a number of updates that supersede items in the specification, particularly a list of the latest CEN test methods relevant to sprayed concrete and a revised section on the execution of spraying.
- *Specification for Tunnelling*, 4th edn. (BTS, 2024).
- ACI 506.2-13: Specification for shotcrete (ACI, 2018).

British Tunnelling Society
ISBN 978-1-83608-693-2
https://doi.org/10.1108/978-1-83608-690-120251003

Chapter 3
Sprayed concrete lining design process and key decisions

3.1. Design process

The SCL design process is a series of steps that develop a number of requirements, through consideration of cost, programme, risk, sustainability and quality, through to a set of outputs, which provide sufficient detail and information to allow tunnel construction to be safely undertaken. Within this process, a number of key decisions should be taken to develop a safe, durable lining system that satisfies the project requirements, and that can be put in place without risk to the workforce, and that complies with the CDM Regulations. This section highlights the sequence of activities involved in the SCL design process and describes the decisions that should be made at each stage.

SCL tunnelling is an open-face method, and the precise nature of ground in which the tunnel is constructed can only be definitely characterised once the face is open. The design process should therefore embody a risk-based approach to decision-making to demonstrate that a robust design has been provided. The design should take sufficient consideration of construction methods, sequences and processes, and make suitable provision for varying site conditions.

The design process should be carried through into the construction phase, because the performance of the lining system needs to be continually verified.

3.2. Key aspects of SCL design

Issues of particular significance to SCL design include the contractual divisions related to design responsibility, the management of risk, and the necessity for integrated construction and buildability input throughout the design. This section details the importance of these aspects to the development of a successful SCL design and provides guidance on best practice.

3.2.1 Progressive design development

The design process involves a series of stages with increasing level of detail (LoD). This guideline considers the typical stages (feasibility/concept, scheme and detailed) through which the design is progressed in a structured manner, adding detail at each stage and ratifying decisions and approaches. Different client organisations and design partners (such as architects using the RIBA workflow) have different titles for each of these stages, but the principles are typically aligned. Through the various stages, the design should be progressively assured, evidencing the verification/ validation of the design outputs for each stage.

3.2.2 Single design responsibility

In SCL design, the distinction between permanent and temporary works can be ambiguous. The commentary offered in BS 6164:2019 clause 6.4.1 should be considered and it should be recognised that temporary ground support provided by sprayed concrete can become part of the permanent works at a later stage.

It is recommended that a single designer is responsible for both temporary and permanent works in so far as they affect the performance of the tunnel lining. If separate designers are employed, a significant interface could be created. This split can lead to duplication, significant interface management effort and ambiguity in responsibilities, leading to the requirement for additional site representation due to the multiple stakeholders involved.

It is, therefore, considered best practice to allocate the responsibility for the design of all lining elements, the conception of temporary measures and for excavation stability to a single entity.

3.2.3   Management of risk

The nature of open face tunnelling, particularly in urban settings, means that a rigorous risk-based design approach needs to be implemented. A high level of attention should be paid to critical areas of the design throughout the design stages, including early reference or conceptual design, so that fundamental risks can be eliminated. The aim is that the risks should be eliminated at source or reduced As Low As Reasonably Practical (ALARP). Constant assessment of risk in terms of health, safety, commercial and programme is essential to ensure that ALARP is achievable and ultimately delivered for the final design.

Robustly designed comparative risk assessments can be a useful tool to demonstrate that the safest and lowest risk option has been selected.

The nature of SCL design means that design decisions and validation continue through the design and construction processes, and the designer should therefore be represented on site and be actively involved in the decision-making/review/approval processes.

3.2.4   Construction input to design

The inherent links between the excavation sequence, face division, cycle time, equipment and construction methodology mean that SCL design and construction cannot be separated, and their interaction is fundamental. Input should be obtained from the construction team intending to undertake the works. This engagement and collaboration should lead to the development of an optimised, safe, robust and efficient design.

Procurement of the design within the framework of the construction contract and resulting responsibilities is critical in terms of demonstrable and effective risk management.

3.2.5   Effective use of analysis

The analytical tools available to the SCL designer are becoming ever more sophisticated. With these advances comes the need for careful consideration of when, by whom, and to what level such analysis should be used during the design process. Analysis should escalate in complexity and sophistication to complement the critical areas of design and LoD required at each stage, as well as the reliability of available information such as the description of the ground.

At the feasibility/concept stage, two-dimensional (2D) analysis is usually sufficient to establish conservative estimates of volumes of excavated material, reinforcement requirements and concrete sections for costing and environmental impact purposes. Precedent experience can assist in determining lining thickening at tunnel junctions or where works are adjacent to existing tunnels or other underground structures. Similarly, for ground movement predictions, empirical or 2D analyses should be undertaken to assess which existing assets are capable of tolerating movements predicted at this stage.

At the detailed stage, analysis of all permanent and temporary workload combinations is needed, together with interaction between the linings and/or waterproofing systems. Three-dimensional (3D) analysis is often employed where third-party asset interaction is significant, or where the geometry/sequencing of the SCL construction is complex.

Across all design stages, SCL design is an iterative process. If the reinforcement quantities, lining stability or ground movements are not acceptable, the lining geometry, construction sequence and face division should be revisited, and the process restarted.

3.2.6   Design coordination and representation

Given the often-complex geometries of SCL, it is well suited to a 3D computer aided design (CAD) environment. The use of building information modelling (BIM) in SCL design is evolving to capture additional information within a 3D (and potentially 4D and 5D) environment. The transition from design to site can be eased by the use of these technologies, which offer the opportunity to improve coordination with other disciplines and the wider design team. All information relevant to design and construction should be available in a format that is readily available, accessible and with a minimum level of ambiguity. Unless other tried and tested technology is available, 2D (PDF) design hardcopy drawings should also be provided.

3.2.7   Best practice guidance/lessons learned

SCL design is an evolutionary process and lessons learned from previous projects should be considered as essential input to determining elements that require critical review during the design process. Knowledgeable design and construction teams bring with them experience, but this can always be supplemented, particularly with respect to specifics of ground conditions and existing asset interaction. If the most relevant information or case studies are not available through publications, previous delivery and client teams should be consulted to identify sources of risk and methods that have been identified to best manage them.

At the end of each project, lessons learned should be gathered and key learning and best practice shared.

3.2.8   Monitoring approach and trigger value derivation

While SCL design in soft ground differs from rock applications (where ground movement might be encouraged/allowed to transfer load into a rock mass rather than the lining), monitoring and responding to ground movements, including their impacts on existing assets, is still a fundamental part of the process. The designer should develop a monitoring scheme that defines limits or triggers for inclusion in action plans for the construction phase. The philosophy behind such trigger levels should be formally recorded and effectively communicated. Refer to Chapter 14: Instrumentation and monitoring.

Where there are other scenarios that require separate triggers – for example, use of compensation grouting in proximity to the lining, adjacent tunnel construction, opening construction and so on – then these should be developed and communicated as separate triggers as part of a specific monitoring plan.

3.2.9   Influence of procurement method on design process

SCL designs can be significantly impacted by the procurement method chosen by the client. The principal implications for design are as follows:

- The later the designer responsible for the design and construction drawings is engaged, the more limited their ability to optimise the design.
- An SCL design taken through to detailed design without the input of the contractor risks rework because excavation sequences are often revised to suit the contractor's means and methods. This can lead to costly and time-consuming re-visiting of the earlier design and analysis stages.
- Contracts that specify separate designers for temporary and permanent works limit opportunities for efficient lining configuration by utilising both primary and secondary linings as permanent works. This arrangement introduces unnecessarily complex design interfaces and assurance processes, creating additional risk.

3.2.10   Sustainability in the design process

SCL tunnels often form part of transportation or utilities projects that, as a whole, provide sustainability benefits to the wider society. Sustainable design is in many ways achieved by providing an efficient design – that is, using less material, which means less impact on the environment, fewer heavy goods vehicle movements, lower carbon footprint and so on. Key opportunities for improving sustainability in SCL design include the following:

- Efficient integration of temporary works and permanent works, particularly using temporary linings as permanent works.
- Optimising excavation volumes through selection of efficient cross-sections.
- Minimising over-excavation.
- Minimising concrete and material usage by minimising lining thicknesses and reducing the use of bar reinforcement through lean designs.
- Specifying concrete delivery methods, mixes and admixtures such that rebound and wastage in concrete lines can be minimised.
- Use of cement-replacement materials.
- Reducing transportation requirements through re-use of excavation arisings.
- Minimising potential for water ingress through the linings, thereby minimising the power requirements of sump pumps.
- Where relevant, specifying surface finishes that reduce lighting requirements.

This list is not in order of importance; it will depend on the specific project and application context. ITA Report No. 35 *Low Carbon Concrete Tunnel Linings* (2024) gives guidance on the relative importance of various factors to the carbon footprint of a concrete lining, which is an important aspect of sustainability but not the only consideration.

3.2.11   Health and safety in the design process

Health and safety is an integral part of design, and should be considered at all stages. The fundamental basis of all SCL design comes down to:

> Can the size and shape of the tunnel be safely delivered using available materials, personnel and equipment?

This question impacts every decision a designer makes, making it of fundamental importance that design teams understand and appreciate the SCL construction process. Consideration of constructability should not be left to contractors alone.

Elements of design that have particular consequence to health and safety include:

- Stability of the excavation
- Risk of contaminated ground, ground gas, unexploded ordnance, encountering existing structures
- Sequencing of the works
- Approach to junction design (the connection of one tunnel to another)
- Joints between sections of the tunnel lining
- Approach to reinforcement.

See Chapter 7 for more details.

3.3. Activities in the feasibility/concept design stage

The first stage of design, known typically as either feasibility, reference or concept design, is driven by high-level requirements and commercial, programme, risk, land availability and geotechnical considerations. The key decisions that should be made during this design stage include the following:

- Confirmation of construction methodology. Determining if SCL is the correct approach for the length, size, worksite availability and ground conditions predicted and workforce health and safety. This is a critical step to demonstrate that the overall project risk is ALARP.
- Implementation of design responsibility split. How any design split and the associated interfaces are managed can be key to delivering an efficient, coordinated design output.
- Requirement for further site investigation (SI) or survey. The designer should determine if additional information is required to reduce uncertainty in design.
- Overall tunnel sequencing. The sequence of tunnel construction in terms of size relationships, drive direction and proximity of tunnels is important and should be decided as early in the design as possible. This should also include the relative proximity of secondary lining construction to excavation and primary lining works.

These activities are fundamental factors that affect health and safety issues, and once fixed have long term implications. Health and safety factors should be considered, in line with CDM regulations, from the outset of the design.

Achieving the appropriate level of confidence in the design typically requires design activities that can be loosely grouped as follows:

- Recording and management of assumptions. At this early point of the design, all information might not be available and any assumptions regarding, for example, the stratigraphy of the ground or the geometry of building foundations should be made. Such assumptions should be recorded and resolved in later design stages.
- Confirmation of requirements. This includes both the client's requirements (such as spaceproofing and alignment) as well as requirements linked to executing the work (work site, materials, staff etc.). The designer should also identify the requirements of additional surveys and site investigations to support further detailed design stages, including those for the resolution of assumptions.

▦ Risk appraisal. Cross-section, alignment and construction sequences should be developed to minimise risk in the construction and operational phases for the workforce as well as the general public to ALARP. Specific information is provided in BS 6164:2019. This is a wide-ranging integrated process that needs to balance a considerable number of often conflicting interests from different stakeholders. The outcome of this process could be the confirmation of whether an SCL tunnel, fulfilling the client's requirements, can be safely constructed and operated in the expected geology without undue impact on other stakeholders (e.g. utility companies, existing structures), the environment or the general public. The risk appraisal should be undertaken in a structured manner and recorded.

The principal outputs from the feasibility/concept design stage include:

▦ Survey and SI scopes
▦ Preliminary ground conditions (from publicly available sources)
▦ Assumptions register and commercial risk and opportunities registers
▦ Initial planning documents – for example, worksite and land take drawings
▦ Preliminary general arrangement drawings
▦ Initial CDM health and safety risk assessments
▦ An estimate of the carbon equivalent emissions as a benchmark for reductions in later design stages. Typically, this would be based on generic values of carbon intensity at this stage.

At this point, the design would typically proceed through a review process and, subject to client approval, then move into the scheme (option development) design stage.

3.4.   Activities in the scheme design stage

Development of the design through an intermediate stage allows the design principles and spaceproofing to be coordinated among design teams and agreed with the client. It must demonstrate that the scheme is aligned among all parties and highlight the key risks. The subsequent detailed design stage can then focus on developing the final design without the risk of abortive work.

A key decision at this design stage is that of determining the lining configuration. SCL tunnels have typically been constructed in two principal phases:

▦ A sprayed concrete primary lining installed as part of the excavation process providing immediate ground support, and
▦ A secondary lining including waterproofing system installation. The secondary lining can be of either sprayed or cast in situ concrete. Where sprayed concrete secondary linings are intended to be used in combination with sheet membranes, international experience suggests that mesh or bar reinforcement should be provided to reduce rebound to an acceptable level.

The primary lining provides immediate ground support. This lining might be able to meet all requirements on its own. The primary lining's long-term structural contribution and load sharing mechanisms with the secondary lining depend on the assumptions made in the design and the choice of waterproofing system.

An overview of potential configurations is presented in Table 3.1.

As indicated in Table 3.1, the choice of lining configuration is closely linked to the waterproofing strategy. Refer to Chapter 9 for further information on designing for water.

Table 3.1 Lining configuration options

Lining configuration	Primary–secondary lining interface	Implications for waterproofing	Implications for division of design responsibility	Implications for sustainable design
Single shell lining	Single lining takes both temporary and permanent loads.	No separate waterproof barrier. Typically considered only where water ingress is likely to be nil/ low, or can be managed within the tunnel.	Single designer.	Potentially optimised lining configuration, but whole life model to include durability/ maintenance considerations.
Double shell lining	Primary lining takes temporary loads, but is assumed as sacrificial in design, so takes no long term/permanent loads. The secondary lining is designed for all permanent loads.	No restriction on choice.	Primary lining – deemed to be temporary works. Secondary lining – permanent works.	Potentially the least efficient approach with respect to lining thickness and materials, depending on durability/maintenance requirements and depth.
Combined shell lining	Primary lining takes temporary loads and a proportion of permanent loads, through a **sharing mechanism** with the secondary lining. Secondary lining takes a proportion of the permanent loads, and typically the full hydrostatic load. No bond or shear capacity assumed between the linings.	No restriction on choice.	Single designer.	Lining thicknesses have the potential to be reduced, increasing efficiency. Waterproofing and secondary lining system choices can be critical to waste and tolerance factors.
Composite shell lining	Primary lining takes temporary loads and a proportion of permanent loads, through **composite action** with the secondary lining. Secondary lining takes a proportion of the permanent loads. Bond (including sufficient shear capacity) assumed between the linings.	Fully bonded waterproof membrane.	Single designer.	Further potential for reducing lining thicknesses. Whole life model to consider preparation required to achieve a fully bonded system and secondary lining finishes.

The developed design should generally build on the previous concept/feasibility stage and consolidate the decisions taken to date. Typically, the baseline geotechnical description for the following design stages should be produced at this point. Likewise, the fundamental approach to the design of the support should be defined and agreed with all parties through a principal document such as a basis of design (BoD), conceptual design statement (CDS) or approval in principle (AIP). This document should also set out the implementation of codes and standards, any deviations from codes or standards and factors/load cases to be assessed plus any other project specific requirements.

Typical activities during scheme design stage include the following:

- Confirmation of tunnel spaceproofing, demonstrating coordination with all interrelated design disciplines.
- Lining and waterproofing strategy appropriate for the performance requirements associated with the structure.
- Identification of durability requirements and a durability assessment.
- Identification of toolbox items on the basis of the risk appraisal and which are appropriate for the proposed excavation and support sequence.
- Progression of the risk appraisal from the feasibility/concept design stage.
- Preliminary excavation sequences based on the strategy developed in the feasibility/concept stage and reflecting the results of initial calculations.
- Initial calculations to obtain indicative lining thickness and reinforcement quantities for the design, as presented at this stage.
- Ground movement analysis. This would be a continuation of the work undertaken in the feasibility/concept stage. During the scheme design stage, a more refined approach is needed, and more attention should be paid to impacted structures and infrastructures.
- Identification of appropriate design codes (Eurocodes/fib Model Code) and headline specifications (BTS *Specification for Tunnelling*) that will be the basis of the detailed design.

The principal outputs from the scheme design stage include:

- Agreed basis of design document (BoD)/CDS/AIP
- 3D CAD/BIM model at LoD appropriate for spatial coordination between disciplines and interfaces
- Drawings indicating tunnel geometry and overall sequencing assumptions
- Updated risk register
- An estimate of the carbon equivalent emissions and a comparison to the benchmark.

At this point the design would typically proceed through a review process and, subject to client approval, then move onto the detailed design stage.

3.5.   Activities in the detailed design stage
Detailed design includes the production of deliverables to be used during construction, the finalisation of risk management and the effective communication and handover of residual risks to the construction team.

Each project is different and it not possible to present a list of activities and decisions that are appropriate for all potential situations. It is, therefore, fundamental to engage with all project stakeholders as early as possible to identify constraints and develop a safe and economic construction approach. Typical activities during detailed design stage include the following:

- Developing the structural and geometrical design to a level where it is ready for issue to the site team. This is an iterative process that establishes the detailed excavation sequencing for individual tunnels, the final geometry and confirms the final lining thicknesses and the reinforcement strategy (plain concrete, bar reinforcement, mesh reinforcement, steel fibres, macro synthetic fibres).
- In collaboration with the site team and the supply chain, developing details for joints, waterproofing, headwalls and secondary lining.
- Identifying specific constraints for the construction methodology, primarily tolerances, maximum advance lengths and minimum and maximum concrete strengths.
- Developing strategies for non-standard situations, such as dealing with known obstructions and the finalisation of toolbox items.
- Establishing quality control criteria, including the monitoring approach and trigger value derivation, and input to pre-construction trials, as well as testing regimes.
- Producing project specific materials and workmanship specifications.
- Progressive reviewing and updating of the risk register. To ensure effective communication and transfer of risk, the contractor should formally acknowledge that they have understood and accepted the residual health and safety design risks.
- Pre-construction trials.
- Contractor's review and acceptance of the SCL design.
- Finalising instrumentation and monitoring (I&M) and establishing trigger values.

The principal outputs from the detailed design stage include:

- 3D CAD/BIM model
- Materials and workmanship specifications
- Construction drawings
- Risk register
- Design check certification including Category 3 check certificates
- An estimate of the carbon equivalent emissions and a comparison to the benchmark.

At this point the design would typically proceed through a final review process and should be subject to final validation to ensure that the detailed design can be taken to the 'issued for construction' status. For further details of design outputs, refer to Chapter 16.

3.5.1 Detailed excavation sequencing for individual tunnels

The designer should develop and agree with the contractor the intended detailed construction sequence (e.g. face division, advance distances and excavation steps). This exercise should optimise the balance between open excavation and productivity for given cross-sections. However, this should be considered along with the prevailing ground conditions, ground movement constraints and the need to close the ring as quickly as possible (for overall stability). Work should continue to optimise the contractor's preferred methods, sequence, plant and equipment for the available workspace.

Construction drawings should be developed and include the following information:

- Advance lengths
- Face divisions

- Ring closure and staggers between any face divisions – for example, the top heading, bench and invert
- Sequencing of adjacent tunnels
- Location of any adjacent structures or utilities and any exclusion zones around them
- Contingency measures to deal with adverse ground or groundwater conditions
- Instrumentation, monitoring and probing requirements.

3.5.2   Limitations on construction

The designer should clearly communicate to the contractor any limitations on construction activities including sequencing, break-outs, proximity of adjacent faces, personnel entry into the excavation face and development of joint details. These constraints should be agreed and coordinated with the contractor.

3.5.3   Design assurance

In parallel with the design, evidence may be required to demonstrate that the design outputs meet the original input requirements. This verification process is not complete until evidence that the original criteria have been met is produced and has been accepted by the client. The exact nature of the assurance process may be dependent on a number of factors including client requirements and standards, procurement methods and design responsibilities.

The key elements of design verification relevant to SCL include the following:

- Evidence that all requirements of the design have been met by specific elements of the design deliverables. For example, design life of 120 years is met by testing detailed in the specification.
- Independent checking of the design deliverables through certification including a fully transparent and auditable process/review documentation.
- The use of peer review or expert panels to review key risks and technical issues.
- Visible mitigation of all health and safety risks to ALARP. Remaining risks reduced to an acceptable level and accepted by the contractor to manage on site.
- Any deviations from codes or standards are accepted by the client.
- Formal acceptance of the design deliverables by the construction team.

Satisfaction of the design assurance validation process should allow the design documents to be 'issued for construction'.

3.6.   Construction phase design input

One of the most important decisions a client needs to make is how the design is managed through the construction phase in terms of the following:

- Undertaking of, and responsibility for, inspecting, assessing, documenting and interpreting the actual ground and water conditions at the tunnel face.
- The procedure for regular review of monitoring data, quality and materials testing data, as-built information (e.g. excavation and SCL profiles), safety incidents (e.g. SCL fallouts or falls of ground) and geotechnical face logs and probing logs. Current practice specifies a shift or daily review meeting should be held, and that the plan for safe tunnel advance for the subsequent shift(s) should be recorded on a Required Excavation and Support Sheet (RESS). This process is described in BS 6164:2019 clause 6.3.3 and in the BTS *Specification for*

Tunnelling, Section 329, and is proven to be an effective method of controlling the safety-critical aspects of the works. The ITA/ITIG Code of Practice for Risk Management of Tunnel Works specifies that 'independent construction supervision' shall lead this process. The independent construction supervision cannot be the contractor or client but could be either an independent organisation or the designer. In either case, the designer of all affected support elements should be represented in this process.

- Defining the key responsibilities for the RESS process, including who compiles the RESS, who can amend it, which parties sign it and what actions to take if the RESS is no longer valid.
- Inspecting, checking and approving the physical works. The designer, along with the client and construction team, should decide which parties need to attend and sign off the works through an inspection and test plan. This process helps ensure that the physical works are constructed according to the design.
- Defining the trials and testing required ahead of the works, and indicating which elements of the specification could be amended post trials, with suitable agreement to optimise the design.
- Indicating which management and testing documents need to be produced by the construction team before works start. The designer should specify the type and number of tests and consider what pre-construction trials are required. Current best practice is to use the BTS *Specification for Tunnelling* as a basis for developing the specification, along with BS 6164:2019 clause 7.7.3.3 respect of minimising dust emissions through mix design.

To achieve the above aims and to ensure a level of independence from the construction team and a clear reporting route to senior management, the construction phase organisation and management structure should be formally documented and communicated. This should include a clear definition of the designer's role and duties within those management arrangements.

emerald
PUBLISHING

ice

British Tunnelling Society
ISBN 978-1-83608-693-2
https://doi.org/10.1108/978-1-83608-690-120251004

Chapter 4
Spaceproofing and alignment

4.1. Spaceproofing

SCL allows flexibility in the geometry of underground infrastructure and this provides the designer the opportunity to optimise the tunnel area and profile to:

- Minimise excavation induced ground movement to limit damage to existing structures and assets, both underground and on the surface
- Reduce excavation volumes thereby minimising construction impact, reducing lorry movements and materials disposal
- Optimise support requirements, allowing reduction in the carbon footprint of construction.

The tunnel geometry is principally governed by the functions that the structure needs to accommodate (e.g. water, sewage, road, metro, rail, high-speed rail, passenger and ventilation tunnels in underground stations, cross-passages, and lift and ventilation shafts) and the associated kinematic envelopes and margins.

The tunnel geometry is often dictated by the performance criteria identified by the client, by an architect tasked with bringing a series of functions into harmony within the tunnel environment, or by the functional needs of other design disciplines (e.g. mechanical, electrical and public health (MEP)/fire/ventilation). Aesthetics, client aspirations and passenger comfort criteria may also influence the tunnel geometry. However, there are additional practical considerations that may have as large an influence on the SCL geometry that should be considered. These include the following:

- Avoiding existing underground obstructions – other tunnels, foundations, services, and so on – especially in urban environments. The size of the tunnel may be limited by rights of properties on the surface or existing underground assets, or the limits for the project set out in the planning process.
- The geotechnical and structural conditions of the ground and the associated loads developing on the structure (the relative magnitude of stresses in the ground), and whether the tunnel takes hydrostatic pressures.
- The equipment and methods used for excavation and construction. It may prove more economical to provide a larger excavation than required to facilitate easier, more efficient and safer construction. Conversely, a flat invert layout is preferred for the excavation logistics, but for tunnels designed under hydrostatic pressure, swelling, or high lateral pressure this may then lead to a requirement for thick and densely reinforced inverts in the final lining. In this instance, a curved invert might ultimately form the optimum solution.
- Where the secondary linings are cast and/or finishes are curved there may be value in standardising profiles to provide uniformity and easy casting of concrete and/or manufacture of finishes. Notwithstanding the flexibility of SCL, the designer should seek consistency and

repeatability within tunnel profiles wherever practicable. Simplifying and rationalising the range of tunnel profiles within a project can provide constructability and production benefits. It is not practicable to deploy a wide range of tunnelling machinery and plant to suit a multitude of different tunnel cross-sections. Long or repeated tunnel profiles make cast in situ secondary linings more cost effective and potentially offer a more sustainable solution with respect to materials, which should be considered as part of the lining configuration decision (Batty *et al.*, 2016).

■ The relationships between different tunnels. At junctions, it facilitates construction if a parent tunnel has a larger diameter than the breakout toward the child tunnel. This can lead to the whole parent tunnel being increased in size or locally enlarged at the junctions alone. These decisions are closely linked to excavation sequencing and the choice of excavation plant. Typically, the ratio between a child tunnel and parent tunnel can be up to about 0.8 before there are significant issues for design and construction.

The flexibility of sprayed concrete can facilitate efficiency in the use of space, particularly with the provision of local cross-sectional enlargements, stub-tunnels and sumps that may only be needed over a short stretch of tunnel. These may be necessary to accommodate additional equipment, hard shoulders and so on. This inherent flexibility may result in a greater number of iterations before the varying demands are balanced, particularly when considering ancillary requirements beyond the main kinematic envelope. For instance, power supply, lighting, ventilation and safety items can often have a degree of flexibility in where they can be located within the underground space. They may be relocated into otherwise unused space in the cross-section that had been created to allow a smooth shape around a given envelope.

As an example, see Figure 4.1. A road tunnel profile is primarily governed by the size of vehicles and number of lanes that the road carries. A typical dual carriageway with hard shoulder results in a rectangular space requirement. Using a round shape, such as that typically associated with a tunnel boring machine (TBM) tunnel, to wrap around this envelope can leave a relatively large volume of space below the road floor level and in the crown. Once selected, a TBM's diameter is typically

Figure 4.1 A comparison between a TBM and SCL cross-section for a typical road tunnel

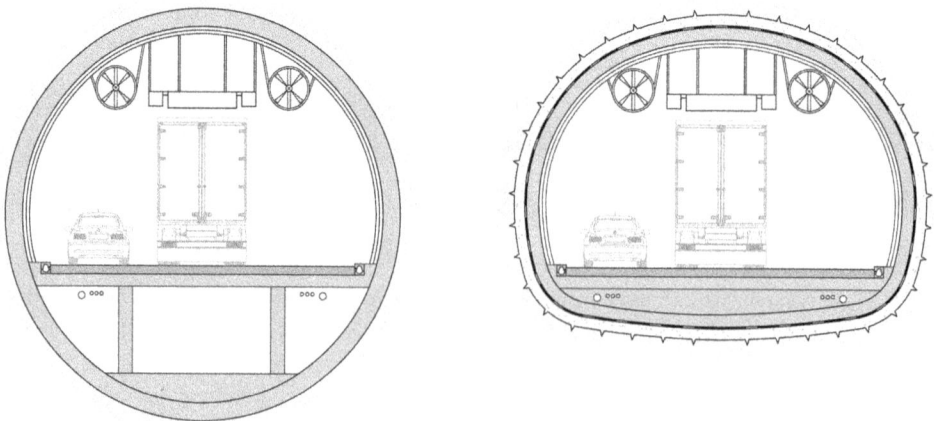

fixed but there being more space than required for drainage, services, communications and ventilation allows for those designs to be finalised later.

By contrast, the use of SCL can dramatically reduce the excavation volumes by using a flatter invert but, in so doing, requires careful spaceproofing of all contributing disciplines at an early stage to optimise the design. Subsequent changes in mechanical, electrical, ventilation, fire or rail systems design may therefore have a significant impact on the SCL design unless this is closely managed.

4.2. Profiles and setting out

Savings in excavation volume can be generated following profile optimisation, improving the flow of stress in the lining and reducing the amount of supporting materials. The reduction in excavation volume may also minimise the impact of tunnelling works on stakeholders and adjacent structures.

Design development offers the opportunity to optimise the profiles to meet the client's spaceproofing aspirations and the contractor's construction needs. Non-circular profiles consist of a series of adjoining curves that avoid stress concentrations, typically a series of circular arcs meeting at common tangents. In soft ground conditions, a flat, large radius invert may need an interim curve to provide the transition to a tighter radius arch. Generally, SCL tunnels are symmetric about the vertical axis, but an asymmetric profile may be chosen to incorporate local features or junctions, thus avoiding wasted excavation on the opposite side of the tunnel.

The designer should anticipate the practicalities of construction by deciding how to define and communicate profile changes to the contractor. 3D models should be considered as they can verify and validate that the design has been communicated correctly. Also, the designer should optimise the number of setting out points, curves and their radii to avoid confusion or errors on site.

The client and designer should discuss buildability and survey control with the contractor at the earliest opportunity. The designer should recognise the practical limitations of the construction process (i.e. the relative accuracy of using a bucket or roadheader for excavation and the application of SCL by way of a spraying nozzle) and the relative precision to which the profiles can be constructed. There is a risk that multiple small changes in radius may be lost and simplified in the field.

Complex junction geometry can be difficult for the designer to specify and for the contractor to construct. Typical surveying methods use an offset from the tunnel centreline to set out the works and monitor the excavation and spraying profiles. These methods do not work well with rapidly changing profiles or headwalls. Therefore, the designer and contractor should agree on how this is best managed on site to ensure the minimum design lining thicknesses are achieved.

4.3. Tolerances

Tolerances should be defined for each component of SCL required for the design (primary lining, secondary lining, water proofing, regulating layers, fire proofing etc.) individually, addressing the items shown in Table 4.1, rather than combined into an overall tolerance.

The contractor should be engaged at the earliest possible stage to decide on appropriate construction tolerances. If a contractor has not been engaged, the designer should state clearly how the tolerances have been developed and how they should be applied.

Table 4.1 Tolerance components

Item	Explanation
Deformation	Should be predicted by the designer who takes account of the ground conditions, excavation profile, minimum and maximum advance lengths, excavation sequence and method of excavation.
Construction	Surveying tolerances, excavation accuracy and sprayed concrete application control limits should be defined by the contractor based on their proposed construction methods. The contractor should also consider excavation equipment, tunnel geometry and operator skills.
Chording	Requires careful consideration if the tunnel is on a curve and a shutter is used to cast a secondary lining.

Lining thicknesses are usually defined during design from the inside of the tunnel to the excavated profile, without including tolerances. The tunnel intrados is the theoretical minimum requirement as coordinated with other disciplines for spaceproofing and tolerances can then be added to the outer dimensions. If the contractor has not been involved in the development of the tolerances, they should be given an opportunity to review these as early as possible. When meeting the client's requirements and the criteria of the specification, the contractor should consider what is reasonable. This approach allows the contractor flexibility to adjust tolerances, and to gain confidence and understanding as tunnelling progresses. However, any changes to tolerances need to be communicated accurately to the survey team (i.e. alterations to the setting out drawings or 3D models). As ever, discipline in version control of drawings or 3D models is crucial and should be managed carefully.

3D models developed by designers should be compatible with the surveying software used by the contractors (i.e. for setting out). The designer should check and assure the models as the contractor may rely on them to execute the works. Going forward, the contractor should be able to amend these models to incorporate the required tolerances.

4.4.   Limitations on curves/inclinations

Depending on the final use of each tunnel, different alignment criteria and other constraints may apply. The BTS *Tunnel Lining Design Guide* (2004) is a good point of reference. The designer should consider the following:

- Expected ground conditions and possible obstructions.
- Land ownership, planning consents, sub-surface acquisition rights, easements and environmental issues (e.g. traffic, noise, wildlife habitats).
- Cover to sub-surface and surface structures, including ground movement, construction, and operational ground-borne noise and vibration impacts.
- Construction planning (worksite locations, size, configuration, geology, logistics set-up and surface access).
- Hydraulic criteria for flow in water tunnels, and reduction in pumping requirements offset against the potential for sediment build-up or stagnant water, laminar or turbulent flow and the implications for spaceproofing.

- Limitations and safe use of SCL plant and equipment, and safety of operatives, particularly on tight radii and steep gradients.
- Practical limitations of SCL surveying techniques (i.e. lines of sight for surveyors), particularly at rapid changes in alignment or profile.
- Limitation of the gradient and vertical curves for vehicles (e.g. typical gradients for rail and road tunnels).
- Limitations on the horizontal curves that can be traversed by the rail rolling stock or lines of sight in road tunnels.

Aside from functional reasons, the designer should consider safety aspects of the gradient and plan the alignment of tunnels either through mitigation of events or their consequences. Factors needing attention in the design of the tunnel alignment include:

- Ventilation and smoke removal
- Drainage criteria, especially for fire events and the removal of extinguishing and flammable liquids
- Passenger/pedestrian escape criteria for tunnels used as evacuation routes
- Minimum curve radii for power cables which might require enlargements at tunnel junctions
- Architectural aspects such as wayfinding and user experience.

emerald PUBLISHING   ice

British Tunnelling Society
ISBN 978-1-83608-693-2
https://doi.org/10.1108/978-1-83608-690-120251005

Chapter 5
Ground and water parameters

The responsibilities for determining ground parameters and the characterisation of the ground should be agreed at an early stage of design. The client may appoint a designer or develop a team to undertake the task of preparing reference conditions. For tunnelling projects, it is essential to gain a good understanding of the ground conditions through which the new infrastructure is to be constructed. As the design matures, the understanding of the ground conditions should evolve. Geotechnical baselines for design and procurement should be produced and included in the contract documentation.

The following selection of publicly available documents provide useful reference information:

- A definition of the responsibility for the characterisation of the ground is given in BS 6164:2019, Section 5.
- The BTS *Tunnel Lining Design Guide* (2004), Section 3, provides a source of reference for geotechnical characterisation.
- BS EN 1997-1: Eurocode 7: Part 1: General rules; and BS EN 1997-2: Eurocode 7: Part 2: Ground investigation and testing; guidance for planning and site investigations.
- Soil and rock descriptions should be carried out in accordance with BS EN ISO 14688-1 and BS EN ISO 14688-2.

5.1. Desk study

A desk study collating the topography, geology, geomorphology, hydrology, hydrogeology and historical uses along the proposed tunnel route should be carried out, and the findings used to develop the ground model. Risks and gaps in information should be identified. A site reconnaissance of the tunnel alignment should be carried out to provide an appreciation of the nature of the alignment and potential shaft and work-site locations. The desk study should also identify obstructions and consider the requirements for the ground investigation.

5.2. Ground investigation

When determining the intensity of in situ ground investigations, the following factors should be reviewed:

- Knowledge of the ground conditions before an initial phase of in situ investigation (i.e. information from desk study).
- Prevalence of surface and sub-surface infrastructure.
- Location of utilities.
- Location of shafts, cross passages, stations and portals.
- Approximate variation of ground and groundwater conditions over the alignment.

- Presence of geological features (e.g. scour hollows and solution features).
- Geological features that could contain high groundwater pressure (e.g. faults, discontinuities and sand lenses).
- Man-made features, such as landfills or mine workings.
- Likelihood of unexploded ordnance (UXO).
- Balance between intrusive and geophysical testing methods.
- Prevalence of strata or deposits that present an elevated risk to tunnelling.
- Location, depth, diameter and spacing of tunnels.
- The risk to existing structures and utilities and the associated need for geotechnical parameters for sophisticated ground constitutive models.
- Type of sampling and in situ testing proposed.

Guidance on the spacing of in situ ground investigations and the frequency of testing is given in Eurocode 7. For linear projects the spacing should be determined depending on the geological nature and the degree of homogeneity of the geotechnical conditions.

5.3.　Ground chemistry

As part of the in situ ground investigation, soil and groundwater testing, along with checks for contamination, should be carried out in accordance with BS EN ISO 14688-1 and BS EN ISO 14688-2. Characterising the aggressivity of the ground and groundwater should inform the selection of a durable concrete mix for the tunnel lining for the proposed design life; see BS EN 1990: Eurocode 0 (for Design life), BS EN 1992-1: Eurocode 2 (for concrete classes and cover to reinforcement) and *BRE Special Digest 1* (BRE Construction Division, 2005).

Any health, safety or environmental risks associated with contaminated ground or hazardous natural chemicals or minerals in the ground or groundwater, must be identified.

5.4.　Ground gas

The soil chemistry and potential for oxidisation should also be assessed and documented as part of the design process.

The ground investigation should also identify the presence of gases in the ground which may be hazardous during construction, or during operation. For more information, refer to BS 6164:2019.

5.5.　Ground parameters

For complex design models, key geotechnical issues include the soil-structure interaction, in situ stresses and stress history, depth-varying parameters (both soil properties and loads), nonlinear stress–strain behaviour and plasticity.

The geotechnical parameters, both for natural ground and areas of ground improvement, should be considered. The parameters used in the analyses should be characteristic values (moderately conservative) as defined in BS EN 1997-1: Eurocode 7.

The variation of parameters should be assessed using a sensitivity analysis for the various geological conditions at locations. The sensitivity analysis can be used to confirm that the design is robust.

Sensitivity analysis should be undertaken on the following underlying geotechnical parameters as a minimum:

- Ground stiffness
- Ground strength
- Ground horizontal to vertical stress ratio
- Ground permeability, in cases where it is not certain whether the ground will exhibit drained or undrained behaviour.

5.6. In situ stress state

The methods used to determine the in situ stress prior to tunnelling should be compared with regional stress information from available structural geological mapping. An important input parameter to establish is the coefficient of earth pressure. The at-rest earth pressure coefficient, K_0, can be established by several laboratory and field techniques.

At-rest earth pressures identified in geotechnical investigation should be used to initialise models. If the data are available, values should be calibrated with regard to recorded lining and ground displacements observed on previous tunnelling projects in the same conditions.

5.7. Groundwater and permeability

The selection of the instrumentation to monitor groundwater pressure should be based on the permeability of the ground. Consideration should be given as to whether the groundwater pressures are beneficial or detrimental to the aspect of the tunnel lining design being considered.

The permeability of the ground influences the rate at which excess pore pressures in the ground dissipate during and following construction and whether short- or long-term parameters govern soil behaviour.

Due account of variations in groundwater levels and the time between acquisition of the monitoring data and the actual construction should be considered.

5.8. Ground stiffness

In situ ground investigations and laboratory testing should be carried out to measure ground stiffness either directly or indirectly.

Stiffness of over-consolidated clays is known to exhibit strongly nonlinear stress–strain behaviour. High stiffness values at very small strains decrease with increasing strain (Jardine *et al.*, 1986). Where predictions of ground movements are important, this small strain behaviour should be considered in numerical modelling using advanced constitutive models.

5.9. Time-dependent ground behaviour

Cohesive soils exhibit both undrained and drained behaviour. Granular material exhibits drained behaviour only. The ability of the ground to generate sufficient passive resistance to maintain stability of the lining should be considered. In low permeability cohesive soils, mobilisation of favourable/unfavourable effects of negative excess pore pressure should be considered.

In the case of clay soils (e.g. London Clay), there may be a gradual increase in ground loading on the tunnel lining as pore pressures change to the long-term equilibrium condition. The time taken for this increase of ground loading depends on groundwater conditions and on ground properties (i.e. soil permeability and consolidation/swelling behaviour of the ground). The construction sequences should be considered in this respect.

5.10. Reporting

Reporting of geotechnical data is described in EN 1997-2: Eurocode 7. This describes two types of geotechnical reports, namely: ground investigation reports (GIRs) and geotechnical design reports (GDRs). Geotechnical baseline reports (GBRs) may be used as part of the contract documents but are not intended as design documents.

At project inception, it is important that the definitions and content of these reports is defined by the client.

Guidance on GBRs is given in ASCE (2007) *Geotechnical Baseline Reports for Construction* and CIRIA guideline (2023) *Geotechnical Baseline Reports: A Guide to Good Practice* (C807).

emerald PUBLISHING ice

British Tunnelling Society
ISBN 978-1-83608-693-2
https://doi.org/10.1108/978-1-83608-690-120251006

Chapter 6
Excavation sequencing

6.1.   General criteria

This section examines the options for the construction sequences that can be used when excavating and supporting tunnels using sprayed concrete in soft ground. The larger the combination of face area and advance length proposed for a tunnel, the greater the risk of instability. This potentially increases the amount of settlement and, in weak ground, the risk of collapse. The ground must therefore, at the very minimum, be capable of being self-supporting for the time it takes to excavate and apply the sprayed concrete along with an allowance for reasonably foreseeable delays.

In many soft ground situations, settlement is related to the distance to ring closure/length of time to close the ring, and is dependent on the properties of the ground to be excavated. How much ground movement is acceptable is therefore a key consideration in the sequence selected. The relative merits of the potential construction sequences need to consider this, as well as achieving a balance of quality, cost and time while maintaining a safe working environment.

The principal options are:

- Excavating full face
- Dividing the face into horizontal strips and excavating heading/invert or top heading/bench/invert
- Dividing the face into vertical strips and excavating side wall drift(s) followed by enlargements
- A combination of vertical and horizontal face divisions
- Excavating pilot tunnels (SCL or segmental lining using a tunnel boring machine (TBM)) followed by enlargement to full profile.

Published guidance on sprayed concrete lined tunnels in soft ground (*Sprayed Concrete Linings (NATM) for tunnels in Soft Ground* – The Institution of Civil Engineers) suggests a face larger than 30 m² should be sub-divided to prevent problems of face instability and settlement. It further suggests that individual vertical heading heights should be limited to 4.5 m without the introduction of benching. These values should be considered as useful starting points, but the design should be finalised based on investigation of the ground conditions, proposed equipment, logistics, detailed analysis and materials. For these reasons, the designer should liaise with the contractor before finalising the construction sequences.

6.2.   Full-face excavation

Excavating and lining a full-face advance in one step allows the lining to be installed as one single element; avoiding joints, which are radial points of potential weakness. Typically, advance lengths in soft ground are limited to 1–1.5 m (see Figure 6.1).

Typical full-face excavation sequence with a sloping face

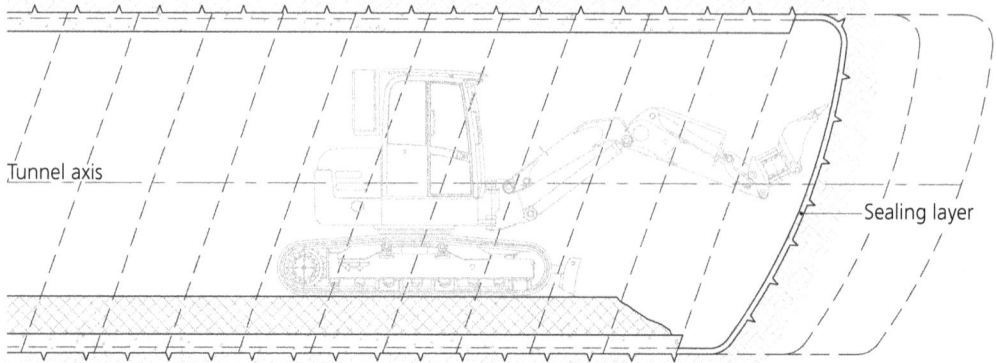

Tunnel axis

Sealing layer

The excavation is not limited by anything but the stability of the final profile and can be bell or circular shaped. With a circular excavation, the tunnel invert may need to be backfilled once the lining is complete to have a flat enough surface for the excavator to work from. This can be done with spoil from the previous advances.

The excavated face may be vertical, domed or inclined forwards, typically by around 20 degrees from vertical.

6.2.1   Considerations

Full-face excavation allows the full lining to be installed in a single sprayed concrete application process, which provides a stable monolithic structure free from radial/longitudinal joints and potential planes of weakness. It also removes the safety risks associated with cleaning and preparing joints.

With this process being relatively simple, the tunnel advance rates can be quick, with very rapid ring closure, limiting settlement. A potential benefit of this is that where compensation grouting is required as part of protection measures for surface structures, the time period when grouting is not permitted (related to the exclusion zone around the open face) can be minimised.

The excavation of a full ring requires an excavator that has a large range of movement. The machine should be able to fit comfortably within the tunnel diameter without conflicting with ventilation and allowing safe access for tunnel operatives to pass the equipment (excavator and spraying robot).

For smaller diameter tunnels, through-mucking excavators can assist with effective spoil removal, as otherwise the process may become too slow for quick ring closure benefits to be realised. In very small tunnels, it may not be possible to use a robot (as illustrated in Figure 6.2) to install the lining and hand spraying may be required. Consideration should be given to whether there are benefits to making the tunnel larger than required in the final condition, to allow increased working space along with greater mechanisation of the excavation and lining installation process.

Figure 6.2 Full-face robotic lining installation (Bank Station Capacity Upgrade)

6.2.2   Where to use

Full-face excavation is an appropriate approach to be used in ground with a sufficient stand-up time to allow for the cross-section to be fully excavated, checked and the lining to be installed with acceptable settlement. This is felt to be viable for most smaller (less than 30m²) tunnels in London Clay, such as metro running tunnels, passageways and many sewers or utility tunnels. The practicalities of this method, particularly the sizing of the equipment (excavator and spraying robot), should be considered.

6.3.   Top heading/bench/invert excavation

Excavating a face in stages of top heading, bench and invert can be carried out in several different ways depending on the ground conditions. The face is divided horizontally into upper and lower parts so that the vertical face height and face area is reduced. This typically results in a roughly semi-circular upper top heading that is excavated in advance of the rest of the face, followed by the lower part, sometimes further divided into bench and invert.

As with full-face excavation, typically in soft ground the top heading advance would be 1–1.5 m. The invert advances are often double the length at 2–3 m and completed behind a series of top headings, as illustrated in Figure 6.3. The spacing between these excavation steps is dependent on the stability of the temporary shape, the ground movement response/limits and tunnel logistics.

Using the previously referenced guidance value of 30 m², a semi-circular tunnel to axis level would have a diameter of around 8 m, although in recent years this has quite often been exceeded without any problems. The height/area of the remainder of the excavation dictates whether it can be excavated in one step or completed in stages with further sub-divisions (bench and invert). As two joints are created for each sub-division of the tunnel, the aim is often to minimise them wherever possible, as they introduce potential sources of poorer quality lining and weakness and – depending on the detailing – may introduce construction stage safety risks. The excavation face may be vertical or inclined forwards.

Figure 6.3 Top heading and invert excavation sequence

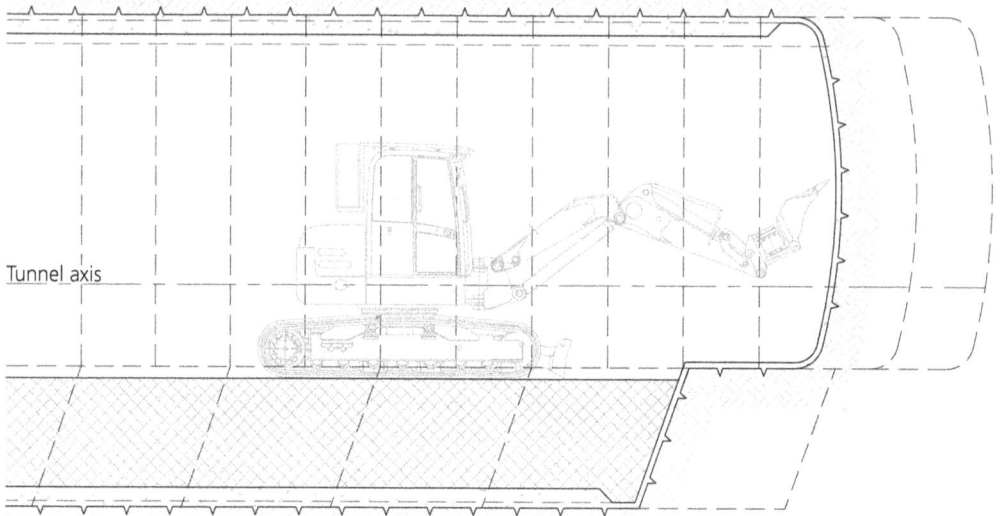

Tunnel axis

6.3.1   Considerations

If a stable top heading can be achieved, then it can be constructed for some distance ahead of the rest of the fully developed face, as demonstrated in Figure 6.4, a photo from the A3 Hindhead Road Tunnel built in soft sandstone. This may require the use of a temporary closed invert or widened footings (colloquially known as 'elephant's feet').

This has advantages of allowing parallel work teams in different parts of the tunnels, one on the bench and one on the heading (subject to safe escape routes being available at all times), and reduces the need for repeated backfilling operations to allow the excavator full reach when switching from an invert excavation and lining stage back to a top heading. While working on multiple faces within the same tunnel offers advantages with respect to time to complete construction, logistics should be very carefully thought through as access to the heading may be severely restricted when work is being carried out on the bench and means to deliver air/water and concrete may have to be doubled up.

In some ground conditions, a temporary sprayed invert strut can be an effective way to temporarily close the ring by helping to distribute localised loads in footings that can cause settlement in the section ahead of full ring closure or prevent movement of the sides of the heading to the centre. In such cases, correct detailing and execution of the connection to the arch is important.

In ground conditions where a flat-bottomed top heading shape is not stable or may not sufficiently halt the progress of ground movements for an extended length, the bench/invert should follow almost immediately, so that a suitable closed ring is achieved as soon as possible. Typically, in London Clay the distance between the furthest top heading and the closed ring is a length equivalent to less than one tunnel diameter. To allow the excavator and spraying robot access to the top heading the lower part of the tunnel often requires backfilling using spoil from recent advances.

Figure 6.4 Top heading and invert separation (A3 Hindhead Road tunnel)

Depending on the length of the heading advance in front of the bench, the reach of the excavator is important if excessive movement of backfill is to be avoided. Ideally, the excavator should be able to excavate the top heading while sitting in the invert with minimal backfilling. Considering the length from start of excavation to achieving closed ring and its impact on stability and ground movement, multiple top heading advances (two to three) may be used followed by the longer bench excavation, as this also reduces the need to move spoil so frequently.

Division of the face in this manner allows a reduction in the face area without the need to use a temporary lining (apart from a temporary top heading invert where required) as all of the applied sprayed concrete gets used within the long-term ground support. However, the joints in the lining, between heading and bench, may be subjected to larger bending moments. It is essential that these radial joints between top heading and bench/invert are correctly detailed and prepared, noting that managing this risk may result in delays in the construction process.

Large tunnels allow for the use of modern large excavators and spraying robots, and the corresponding economies of scale. For example, it may be possible to use a loading shovel and excavator, which might allow for a faster excavation process, and a larger robotic spraying robot with a faster pump rate.

6.3.2   Where to use

Top heading, bench and invert construction can be used in tunnels where a larger tunnel is required that can be effectively excavated and lined in a single span, subject to the ground not being prone to regular large wedge failures relative to the heading width. Settlement control may be achieved by completing the bench and invert sequence to provide full profile in as short a length as possible. Alternatively, settlement may be controlled by constructing a temporary invert strut in the top heading sequence, allowing the contractor to construct on two faces without the need for constant backfilling between top heading and bench/invert.

6.4. Sidewall drift excavation

Sidewall drift construction involves splitting the face vertically. Depending on the size of the tunnel and ground conditions, single or double sidewall drifts are used. To create a suitable shape for stress redistribution and prevention from buckling of the temporary sidewalls in the temporary condition these are often initial cathedral shaped excavations followed by a crescent or mushroom shaped excavation to complete to profile. The temporary walls can be left in place during excavation of the remainder of the face and only removed once the full profile is completed, providing support and potentially reducing ground movement impacts.

As with the top heading/bench/invert divisions each time the face is divided a joint and potential location of weakness is introduced, so the smallest number of divisions possible should be used. In recent years, in soft ground, single sidewall drifts have been used for tunnels with spans between 11–15 m,(see Figure 6.5) with double sidewall drift methods used for tunnels with spans over 15 m (see Figure 6.6).

Figure 6.5 Single sidewall drift excavation (Tottenham Court Road Station Upgrade)

6.4.1 Considerations

Sidewall drifts offer an alternative way of reducing the area of open excavation, with a potential advantage of allowing a closed ring in the temporary case so that ground movements can be limited. The sidewall drift can be advanced ahead of the enlargement, allowing parallel progress as with the horizontal divisions, but with different implications for access/egress and logistics. Both the non-circular geometry and limits on sidewall drift width, narrow the choice and efficiency of the excavator and spraying robot that can be used, with this impact increasing with smaller tunnels. Coupled with more complex setting out this can make sidewall profiles slower to complete,

Figure 6.6 Double sidewall drift excavation (Crossrail Stepney Green cavern)

although this might be balanced by the ground movement limitation benefits offered by the closed shape. Using the sidewall drift method, the majority of the lining can remain a component of the long-term ground support.

The stability of the completed profile is reliant on well designed and executed joints, particularly those in the crown and invert where temporary sidewalls meet the main structural lining. The joints are more complex than those typically found in a horizontally divided face, in part due to the retention of the temporary wall during the enlargement. Demolition of the temporary sidewalls and invert struts, particularly near the connection zone to enlargement lining requires care not to diminish structural integrity of joints, or the durability of the long-term ground support.

The sidewall drifts can be further divided horizontally into upper and lower sub-headings.

Due to the complexity of the joints and shapes, sidewall drifts are often reinforced or use lattice girders to guide the excavation profile. This requires more lifting and fixing and work in the tunnel face, often at height, than alternative excavation and support sequences considered here. It is current UK practice to design out lattice girders wherever possible.

6.4.2   Where to use

Sidewall drifts are advantageous where tunnels are particularly wide, or increase in width (e.g. rail turnouts and crossovers). Below a certain face area, they offer limited benefits and are impractical to build when compared with a horizontal split into top heading, bench and invert. Where there are risks of potential ground instability, sidewall drifts help sub-divide the face and reduce the likelihood and consequences.

6.5.   Pilot tunnel enlargement

Pilot tunnel enlargement construction is typically used in soft ground where the face area of a tunnel to be excavated is too large to be constructed comfortably using a top heading/bench and invert or there are potential face stability issues. The method offers simpler construction sequences and processes compared with the sidewall drift method, with less complicated joints coupled with simple ring closure.

The pilot tunnel enlargement method involves the initial construction of a smaller diameter tunnel that is later broken out/enlarged to a full-size profile. The pilot tunnel can be constructed from SCL (see Figure 6.7) or a precast concrete lining that could be constructed using a TBM. The TBM could be continuing with excavation of other tunnel profiles in the project, such as a running tunnel that is to be locally enlarged to a platform tunnel profile (see Figure 6.8).

Figure 6.7 SCL pilot tunnel in a top heading invert enlargement (Bank Station Capacity Upgrade)

Limitations on the pilot size are similar to those for any other smaller face excavation. The enlargement is typically excavated in a top heading/bench/invert sequence, but with the advantage that the face area in one advance is reduced, with any potential face wedges being supported/broken up by the pilot tunnel lining. The pilot tunnel lining also potentially limits settlement by restraining movement of ground during the enlargement excavation towards the face.

The pilot tunnel offers the opportunity to improve knowledge of ground and water conditions, allowing for mitigation measures to be planned for the larger span and possible ground treatment and de-watering, and so on, installed.

6.5.1   Considerations

Pilot tunnels constructed using SCL potentially require the project to have a range of tunnelling equipment available. Pilot tunnels have similar equipment and logistics limitations to other full/smaller face excavations, while the enlargements typically require the use of large excavators and spray robots to achieve the reach required.

Figure 6.8 Segmental pilot tunnel in a top heading invert enlargement (Crossrail, Tottenham Court Road Station)

Pilot tunnels built using TBMs that are constructing other tunnels on the project may have the advantage of allowing the TBM to progress without having to be transferred or moved. Typically, it would be necessary to wait for the TBM's activities to finish before starting the enlargement, unless there are parallel tunnels that can share logistics and muck-away behind a certain point.

The pilot tunnel size may impact the speed of the enlargement sequence and influence the development of ground movement. If the pilot tunnel is too small compared with the final profile, some benefits in reducing the enlargement face may not be realised. It should be noted that the pilot tunnel can be a significant tunnel structure in its own right (BS 6164:2019, cl 5.4.3).

A drawback of a larger pilot is the greater amount of lost/removed material (concrete) in the pilot, with associated costs and environmental impacts, as well as greater initial settlement and a higher risk of face collapse (particularly if the pilot is being used to identify or treat weaker ground).

If a pilot tunnel is being used in conjunction with other ground movement control mechanisms, such as compensation grouting or ground treatment, there are further considerations:

- A circular pilot tunnel is more stable under dynamic loads from compensation grouting than a flatter shape.
- A larger pilot tunnel might avoid the need for repeated grouting processes, if settlement induced by the pilot tunnel is a large proportion of the total settlement. Thus, compensation grouting could become a one-time only operation, significantly saving on the cycle time of the enlargement.
- A pilot tunnel being used to allow installation of ground treatment or drainage would need to be sufficiently large to allow the equipment to drill out from potentially both the face and the side walls (see Figure 6.9).

Figure 6.9 Probing from inside a segmental pilot to facilitate an SCL enlargement (Crossrail, Farringdon Station)

Figure 6.9 Probing from inside a segmental pilot to facilitate an SCL enlargement (Crossrail, Farringdon Station)

The position of the pilot tunnel relative to the enlarged tunnel can vary, from above the crown to at the invert. The decision on the pilot tunnel location is influenced by the following:

- The complexity of joints and whether parts of the pilot are to be incorporated into the permanent lining. If the pilot is centrally located and completely removed, joints in the enlargement between the heading, bench and invert are relatively simple, making them easier to construct and, therefore, structurally safer and more robust.
- Whether excavation of the enlargement might be hindered by the position of the pilot tunnel within the cross section. If it is difficult to get an excavator around the pilot tunnel, this might increase the time to excavate and install the lining.
- Whether the pilot tunnel is being used to install ground treatment/drainage and what location allows optimal coverage for the enlargement.
- Whether there are concerns about crown stability of the enlargement. Placing the pilot at or above the crown of the enlargement (see Figure 6.10) can provide additional safety to the tunnel crew, as they are working beneath a pre-installed roof canopy, although this may be offset by a greater requirement to work in the face to create joints.
- The impact on ground stabilisation and ground movement should be considered when determining the location of the pilot tunnel, whether it be at or above the crown or in the invert of the enlarged tunnel.

Access to the area of construction of the enlarged tunnel could be from within the smaller tunnel or from an interim point by breaking into the precast concrete segments from the side. The second option could be facilitated by building a tunnel in advance, perpendicular to the TBM alignment, which is backfilled with an engineered fill (such as foam or entrained concrete) before the TBM passes through, as illustrated in Figure 6.11. By this means, a stable/controlled zone is created

Figure 6.10 Pilot tunnel in the crown of a tunnel (Kings Cross Station upgrade)

Figure 6.11 Connection into TBM Pilot Tunnel from a backfilled SCL wraparound (Crossrail Tottenham Court Road Station)

where removal of the pilot tunnels segments can take place without the risk of ground instability/structural collapse.

When starting the enlargement from within the smaller tunnel, it is normal practice to transition to the larger profile over a number of advances. The enlargement is started by removing a section of the pilot tunnel and installing a lining that flares from the pilot tunnel profile to a slightly larger profile, so that at the end of the new section the tunnel lining is offset from the pilot tunnel ahead. On breaking out the next ring a slightly larger tunnel is built, again flaring gradually up to a larger profile. Each advance may need to be divided into top heading, bench and invert as the profile gets larger. The angle of the flare and number of advances in the transition depends upon the difference between the geometry of the two sections and the practicalities of installing the lining. The process is similar for SCL (see Figure 6.12) and precast concrete segmental pilot tunnels.

Figure 6.12 Enlarging from a pilot to the full cross-section

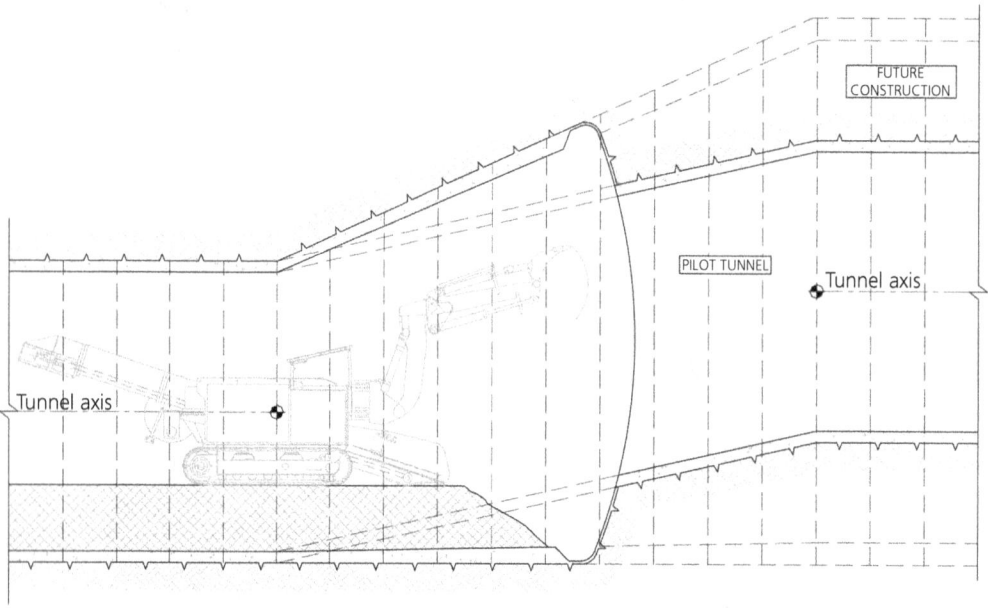

With an SCL pilot tunnel there is a higher risk of causing collapse of the lining some way along the pilot tunnel before the enlargement lining is in place, if not pre-cut into shorter sections. With a segmental lining there may be a need to cut dowels bars or remove bolts ahead of dismantling the pilot tunnel. During dismantling, this can make a precast concrete segmental ring less stable than a partial SCL, with some circumferential and longitudinal connectivity. A critical success driver for the pilot tunnel enlargement method is a design that facilitates the fast, dust-free removal of the pilot tunnel.

On the Crossrail project, precast concrete segmental linings were easily removed. Using an excavator the segments can be pulled down to be broken up away from the face and removed with other spoil. Experiences with SCL pilot tunnels have been varied, with some removal processes proving to be slow, laborious and dusty, generating high levels of respirable crystalline silica. Ease

Figure 6.13 Excavator attachment used to break out a pilot tunnel (Bond Street Station Upgrade)

Figure 6.13 Excavator attachment used to break out a pilot tunnel (Bond Street Station Upgrade)

of removal may depend upon the equipment available for the enlargement, with munching (see Figure 6.13) proving preferential over grinding methods. Some SCL designs have attempted to pre-cut the lining, using a de-bonding material or mechanically applied plastic or steel arches.

6.5.2 Where to use

It is expected that, unless for other logistical reasons, pilots will mostly be used in tunnels with diameters greater than 9 m, where there are concerns about face stability using a top heading, bench and invert sequence or where methods to control settlement are proposed, such a compensation grouting.

Pilot tunnels may also be used to identify in advance features that require pre-treatment before the full excavation, particularly where a full ground investigation from surface has not been possible or questions remain about certain features and how they will respond to excavation.

6.6. Vertical and inclined excavations

SCL construction is frequently used for shaft and inclined tunnel excavations such as escalator tunnels. Typically, the upper part of a shaft, through made ground, fill and sands is constructed using caisson sinking techniques, diaphragm walls or piles. SCL then takes over once stable ground has been reached.

6.6.1 Vertical shafts

Vertical shaft construction in SCL is fairly straightforward. Depending on the size of the shaft, they are either constructed in single full-face sections or broken into quadrants. The depth for each excavation stage is typically 1 m. Joints between sections should be detailed as with any other circumferential joint to provide good continuity.

6.6.2 Escalators

Historically, inclined excavations for escalators were constructed downwards. They tend to have large face areas (particularly the machine rooms) and have an inclination of 30 degrees. They are usually built with the face perpendicular to the inclined centreline, so it is therefore angled away from the workforce. The face is typically divided in a similar way to horizontal tunnels (i.e. dividing the face into top heading, bench and invert or using a pilot tunnel). Plant should be capable of working on steep slopes and, in some cases, be specially secured. Consideration should be given to developing efficient methods for removing excavated material, such as creating space for a conveyor system. Refer to BS 6164:2019, clause 8.5.2.

In recent years, upward escalator construction has been used to meet programme and logistical requirements where projects have complex interfaces between construction sites and methods (e.g. boxes and tunnels built by different contractors). One example of this construction method was on the Crossrail project at Liverpool Street Station where specialised plant, including overhead rail supported systems, were used.

When tunnelling upwards, and to meet health and safety restrictions, highly mechanised, special excavation and support equipment has been employed to prevent plant moving down the slope. Rail-mounted excavation and sprayed concrete units, consisting of a mainframe that is hung from the crown and travelling along parallel rails, have been developed. This required local strengthening of the primary lining. Alternatively, a sequence could be developed where the excavator is effectively flat at each stage but excavating the top heading upwards, similarly to when enlarging the profile. This approach requires careful consideration of backfill and possibly use of temporary inverts, allowing the full invert to be excavated downhill.

British Tunnelling Society
ISBN 978-1-83608-693-2
https://doi.org/10.1108/978-1-83608-690-120251008

Chapter 7
Design details

As design progresses, effort will transfer from spaceproofing and coordination with other design disciplines onto designing and communicating details related specifically to the tunnel construction. However, these items should have been given some consideration in early decision making related to tunnel geometry and excavation and support sequencing, as they are inherent to the safety of the proposed design. These key items can include:

- Joints
- Junction design
- Reinforcement strategy
- Probing
- Temporary support measures for face stability
- Ground improvement.

7.1.   Joints

Joints in tunnels are divided into two groups:

- Circumferential joints (also known as circle joints), are joints that run around a tunnel's circumference. In a segmental tunnel, it would be the joint between full rings. In SCL, it is the joint between advances, which may not align between the top heading, bench and invert.
- Radial joints (also known as longitudinal joints or seat joints) are joints that connect different sections of a tunnel ring. A full-face excavation and lining would have no radial joints.

Discontinuities in the circumferential direction are not significant for lining stability, provided that each 'ring' is working as the design intended. At the end of an advance, it is good practice to leave a chamfered leading edge at the face because it provides the correct profile for the next advance and spraying as it builds up the section to full thickness. Provided that the joints are clean before the application of the next layer, there are few quality concerns.

In some areas, particularly for thicker sections and where lattice girders are used, the sprayed sections are overlapped – for example, the last 75 mm of the previous advance is sprayed at the same time as the initial layer of the next advance. This does have implications for the control of the exclusion zone close to the excavated face.

Radial joint design requires additional consideration by the designer because the joints should have a bond, and not include rebound or other debris, to allow effective load transfer. The ability to achieve this with the geometry proposed will be critical to the successful installation of the lining.

To address this issue of joint details, any preparation/methodology should be prescribed in the specification and/or drawings. There are three methods that have been used in recent years:

- Connection with reinforcing bars, using a starter bar system that can be pulled down when the underside is exposed, or post drilled from below (see Figure 7.1).
- Using elephant's feet at the bottom of the top heading.
- Building up the lining across the joint in layers, so that there is not a single distinct zone of discontinuity.

Figure 7.1 Joint continuity bars (Tottenham Court Road Station Upgrade)

During the Crossrail project, it became standard to include a continuity strip that reinforced the radial joints as per the first bullet. A mechanical connection across the joint and a rough surface on the underside allowed the subsequent section of lining to key to the upper section. However, installation of this element required personnel entry at the face prior to spraying the top heading primary lining and again after bench/invert excavation to pull down the bars, increasing risk to the workforce from falls of ground or fresh shotcrete sprayed concrete. Determination of time limits and concrete early-age strength criteria to allow access to the face requires trials and testing, and close collaboration between the construction and design teams, and rigorous application on site. Further details of exclusion zones, which must be vigorously managed using this approach, are provided in the Crossrail Best Practice Guide, SCL Exclusion Zone Management.

Where access is unavoidable, a comparative risk assessment of the joint quality and design robustness should demonstrate that the chosen solution has reduced the residual risk to ALARP (St John *et al.*, 2016).

As well as the safety aspects of reinforced joints, there have also been difficulties in excavating the lower advance without causing damage to the bars. This may make it harder to get good cover, spray the around bars and get good encapsulation to achieve high quality concrete at the joints.

An elephant's foot at the base of an excavation involves widening of the lining to the outside of the profile. With competent foundations (not in soft ground) this could be used as the permanent foundation. In soft ground this shape can spread the load of the first stages of excavation until the ring is subsequently closed, thereby providing stability that can be considered in the exclusion zone management.

Since Crossrail, the trend has been to, wherever possible, remove the need for hand work near to open ground. Figure 7.2 shows an example of a stepped joint that avoids the need for personnel to enter the face. Here, the joint is built up in layers, similar to the longitudinal joints using a leading edge.

Figure 7.2 Stepped joint example

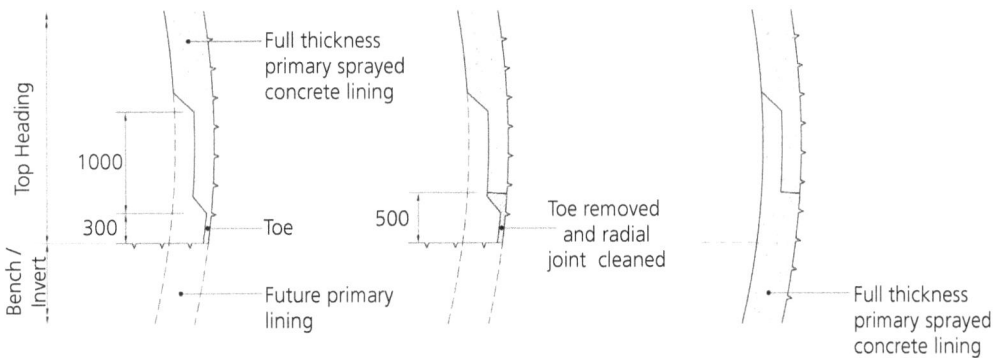

This approach has been successfully applied on the Bank Station Capacity Upgrade and in SCL works on HS2. Care must be taken to follow the steps and ensure that the joints are properly cleaned before subsequent layers are applied. Coring of joint samples to check there is not visible delineation between layers is recommended. Benefits include simplifying exclusion zone management, as there should be no need for anyone to enter the face in a typical excavation and support cycle.

This solution will not work in all joint arrangements, notably where the load path is not continuous. In all cases, the joints should be properly reviewed and signed off before tunnelling continues. Design specifications should include an approach to joint rectification in the case that supervision is not satisfied with the quality at an interim point.

7.2.   Approach to junction design

Openings from one SCL structure (parent tunnel) to start construction of another SCL structure (child tunnel) may not always provide the latter with a fully closed SCL tunnel structure for the first few advances, and some areas of concrete may be overhanging. Available measures to deal with this situation include the following:

- Steel bar reinforcement (e.g. L-bars) at connections between parent and child SCL tunnels, where more than the first child tunnel advance is affected.

45

- Steel bar reinforcement and local widening (e.g. elephant's foot) of the SCL, where the overhang is exceptionally large.

Current UK practice for the design of the child/parent tunnel junction is to thicken the parent tunnel lining profile on either side of the future child tunnel. The child tunnel lining is not thickened as the parent lining is designed for the opening. The lining of the parent tunnel within the eye of the opening is usually not thickened because this will be broken out. Other options are available such as the provision of a thickened ring around the eye, ring beams either side of the eye and/or a thickening of the child tunnel adjacent to the opening.

Fast ring closure of the child tunnel should be carried out to frame the opening created in the parent tunnel. Where L-bars or other connections are specified, their compatibility with the excavation sequence advance length of the child tunnel should be considered.

7.3. Reinforcement strategy

Bar reinforcement should be avoided where possible in sprayed concrete linings for the following reasons:

- Spraying around bars has practical, durability and quality implications due to shadowing and voiding.
- It is difficult to install where reinforcement continuity is required.
- The installation of bar reinforcement presents safety issues as work is often required in the crown and shoulders of the tunnel (working at height, proximity to excavation face and under fresh sprayed concrete etc.).

Fibre reinforced concrete is therefore current best practice for both primary linings and addressing serviceability of sprayed secondary linings (for thermal and shrinkage issues).

Consideration should primarily be given to the shape and thickness of the lining to minimise the need for bar reinforcement to support the fibres. In many cases, stress concentrations can be addressed through tweaks to geometry. The locations where this can be difficult include at junctions to other tunnels and in inverts where a flat shape might be unavoidable due to other constraints.

Where reinforcement around an opening is required in the primary lining, it may be possible to install it in a subsequent lining thickening. This is a layer added after the main excavation of the tunnel drive is complete, prior to creating the opening. This was used on the Crossrail project, allowing bars/mesh to be installed in a safe environment. However, this can lead to very thick linings at openings and it might be that a different approach to the linings and analysis can reduce the requirements more efficiently.

7.4. Probing ahead

Probing ahead is required to assess the nature of the ground conditions ahead of the face or adjacent to the tunnel. This allows corroboration of the ground model as understood during the design and provides additional information on features that could not be fully described by ground investigation from the surface. This, in turn, informs application of toolbox items sufficiently in advance to be most effective such as the application of drainage or ground treatment.

Probe drilling is simple, open holed, non-instrumented drilling ahead of the excavation face. It is normally carried out with three to five probe holes, depending on tunnel size, within the top heading or pilot face, usually in slightly varying directions. While this is still a limited array, it can pick up potential obstructions, or provide better information on the extent of non-continuous geological features not fully described by the ground investigation, such as sand channels or hard bands.

During the detailed design the number and location of probe holes should be specified. Where there are features of particular concern, the number and direction of probe holes should be adjusted to address the risk.

A systematic record of probe-hole positions, the nature of the ground disclosed, drill penetration rates and water ingress should be recorded and fed back to the designer during the construction process

7.5. Toolbox items

SCL construction allows for the use of various specific tunnelling techniques, sometimes referred to as toolbox items, to be used in either a systematic or a non-systematic way. The installation of these toolbox items depends on local ground conditions and while a design may be developed in advance, selection and installation would often be based on site observations; made by the engineers/geologists acting as the designer's representative on site or directly on behalf of the contractor. The likely effect of toolbox items should be determined in advance and confirmed by monitoring once in use.

The content of the toolbox should be assessed prior to tunnelling and continually reviewed as the project progresses. Its use should be proposed and accepted at the on-site daily review meetings. Resources, equipment, and materials should be readily available to allow the toolbox items to be immediately implemented when they are required. Toolbox items should be repeatedly applied until the tunnel progresses beyond the area of concern. Potential toolbox measures are described below.

7.5.1 Supporting the face

A number of methods can be used to make a face more stable at an interim excavation step. These can be applied systematically throughout, or in response to worsening ground conditions or approaching a known feature. They include:

- Sealing with sprayed concrete
- Berms
- Sloping/angling the face.

Face sealing involves spraying concrete onto the face of the excavation rather than just onto the walls. It seals the face, preventing it from drying out and properties deteriorating before the next excavation step. It can also prevent ravelling of sand, or flowing sand due to groundwater seepage, as long as the seepage is not so bad that you cannot spray onto it. Also, in very poor ground, if the face is divided and face nails are used, then the face the sealing layer will provide support between the face nails and enable the use of face plates and nuts to mobilise the nail better. As the sealing layer will be removed with the subsequent excavation step, it should be kept to a minimum thickness required to avoid material wastage. If there is a need for personnel to enter near the face, then a sealing layer will reduce the risk of block falls, but might need to be thicker, to provide sufficient structural capacity for this task.

The centre of the face can be supported using material from previous excavations, applied in a mound or berm, or left in place during the excavation of the outer edges. This can increase standup time of the tunnel face, but must be sized so as not to impact the excavation and spraying processes. This can be used in combination with pocket excavations, described shortly.

The tunnel face can be sloped or domed to increase stability, or reduce the risks to the workforce of face collapse. Similarly to the berm, they can allow the support to the tunnel lining to be installed without the centre of the face being required to stand in a vertical plane.

7.5.2   Reduced advance lengths

While excavation advance lengths in soft ground are typically 1 m in length, these can be reduced from their normal value if there are stability/settlement concerns that might be addressed through a shorter excavation and support cycle. It is noted that this may not reduce the face profile and in weaker ground the advance length may be hard to control, due to fall outs and so forth.

7.5.3   Pocket excavations/face splitting

Smaller areas (pockets) of ground can be excavated and immediately sealed with an initial layer, to prevent the ground drying out or to quickly stabilise it. This could be through splitting the face in two, into strips, or the pocket excavation can be applied as required to limited areas in the tunnel face. The pockets could follow the circumference of the excavation, leaving the centre of the tunnel supported using spoil from a previous excavation or simply remaining sealed from the last advance.

In order that the lining is not compromised by additional numerous joints the sealing layer should be applied leaving sufficient room to install the primary sprayed concrete lining, particularly if it is part of the permanent lining.

7.5.4   Spiles

Spiles are steel or glass reinforced plastic (GRP) rods inserted around a tunnel profile. They should be supported by the tunnel lining (often with the support of lattice girders, as illustrated in Figure 7.3) at one end and embedded beyond the excavation advance at the other. They support the excavation profile by reducing the span over which the ground must support itself. They are often installed raked at a slight upwards or outwards angle, so that further overlapping arrays can be built inside the protected profile in subsequent advances. Typical installation lengths are generally 3–4 m. Due to their short length, spiles are usually installed every advance or every other advance. The limiting factor is often the need for overlap while ensuring there is sufficient embedment ahead of the face. The longer the spiles are without a further array installed inside, the larger the potential profile if the ground breaks or ravels back to the line of spiles.

Spiles are typically used in the crown and, depending on the geotechnical hazards to be mitigated, will be installed over an arc of 90–180 degrees. They can range from standard reinforcement bars rammed into the ground to purpose-made self-drilling spiles that are hollow to allow grouting of the ground around them. Note that spiles will do very little to improve face stability, they only help protect the profile of the excavation. It should also be noted that their capacity is reliant on the face being stable. If the face is unstable, then the spiles do not have any support at their distal end and will be cantilevering from their supported end in the installed lining. In very poor ground, trench sheets have been driven into the ground, rather than discrete spiles.

Figure 7.3 Spiles being installed with lattice girders (Crossrail Stepney Green Cavern)

They are not typically considered in the permanent case. If used with a permanent primary lining, they will often be installed in an enlargement to the normal profile.

7.5.5   Pipe arches

A pipe arch forms a canopy around the excavation profile in a similar way to spiles but in this case using steel pipe of typically 114–168 mm diameter. These have with greater stiffness and higher capacity in bending and are typically installed over longer lengths; about 9–18 m. As well as inhibiting localised ground failure in the crown, pipe arches have been used to limit ground movement impacts on assets in close proximity to the tunnel by providing a stiff canopy that shields the adjacent asset (Farrell and Terry, 2015).

If only one set is being installed, they may be able to be installed parallel to the tunnel centreline, allowing for installation tolerances. If used in a sequence over a length longer than one set of tubes, the canopy tubes are installed fanning radially outwards, to allow space for the next set to be installed inside them while remaining outside the tunnel lining profile. See Figure 7.4, where pipe arch canopies were installed in a stepped umbrella array at the south portal of the A3 Hindhead Tunnel.

The layout of pipe arches should take account of the following:

- The positions of services as shown on all available services and survey drawings, as well as any uncovered during the initial stages of execution of installation/grouting.
- Any restrictions relating to temporary or permanent works that the execution of the pipe arch installation may affect.

- All restrictions on access and sequence related to traffic management, environmental requirements and commitments to third parties.
- Setting out tolerances.
- Deviations from theoretical drilling axis.
- The potential that they can form a plane of weakness in the ground resulting in the ground locally falling out and back to the line of the pipe arch/spile during excavation.

All installed steel pipes should be checked and verified as to their correct location prior to being grouted. Any steel pipe deviating by more than the specified tolerance should be replaced by an additional pipe.

The worst design case will be just before drilling the next round of canopy tubes, when the length of the pipes are at their shortest and the excavation cross-section is at its largest. The shortest length is known as the overlap length, and will be typically 6–8 m.

Similarly to spiles, if the distal end is not supported, the capacity of the pipe arch will be more limited. Therefore, canopy tubes might be used in conjunction with face division and/or face dowels, and sometimes also ground treatment.

7.5.6   Face dowels

Face dowels are grouted rods installed into the tunnel face that may prevent circular slip failures or block failures by holding the ground ahead together. Like spiles, they can be steel or GRP, and can be solid or hollow. However, unlike spiles, they are excavated as the tunnel advances, so consideration of their removal is needed.

Face dowels are installed when calculations or observations show the face is not stable. Therefore, they are only used in poor ground, or when it is necessary to open up large faces.

Face dowels work in a similar manner to soil nails used to support slopes. Assuming the dowel itself does not fail in tension, then for any potential failure mechanism, either the ground fails around the dowel, or the dowel is pulled out of the stable ground beyond the failing ground mass. Assuming the bond strength is the same in both zones of ground, then the shorter distance will determine the pull-out capacity. For this reason, it is essential that dowels always extend sufficient distance beyond the potential failure zone into stable ground. The worst design case will be just before drilling the next round of face dowels, when the length of the dowel is at its shortest. This is known as the overlap length, and will be typically 6–12 m.

Anecdotal experience suggests that they were not proven to be beneficial at Heathrow Terminal 4 Station (London Clay), Crossrail (London Clay) or Hindhead (weak sandstone). At Heathrow Terminal 4 Station and Crossrail, the face dowels did not noticeably reduce ground movements. At Hindhead, it seemed that the disturbance caused by drilling cancelled out any beneficial effect once the dowels were installed.

7.5.7   Local dewatering/depressurisation

Local ground depressurisation to reduce porewater pressures can be used as a toolbox item. While performance requirements will be specified by the designer, it is typically detailed by specialists.

7.5.8   Ground treatment

Typically, SCL construction in London Clay or other stable ground conditions can be undertaken without the need for any ground treatment. In areas of poor ground, water-bearing soils and non-cohesive ground conditions, or if there is low cover, a range of ground treatment measures can be implemented. These include:

- Permeation grouting
- Compaction grouting
- Compensation grouting
- Jet grouting
- Deep soil mixing
- Ground freezing.

When considering these solutions, the stability of non-circular profiles needs to be carefully considered in the design as not all shapes would be stable under the dynamic loads induced by grouting or ground freezing.

It is common to dictate zones where ground treatment should not be actively installed where this may result in additional loading on unsupported ground and while sprayed concrete is developing its early strength. This is mainly for compensation grouting but can also apply to late introduction of other grouting or freezing operations. As with depressurisation, often the SCL designer is specifying performance requirements and specialists will provide expert design input. If such measures are anticipated, it is important that the ground investigation provides sufficient details for these items to be designed properly.

emerald PUBLISHING   ice

British Tunnelling Society
ISBN 978-1-83608-693-2
https://doi.org/10.1108/978-1-83608-690-120251009
Emerald Publishing Limited: All rights reserved

Chapter 8
Material parameters

8.1. Sprayed concrete

Sprayed concrete consists of the same constituent materials as cast in situ concrete. The general material behaviour of sprayed concrete follows the same set of rules as defined in existing concrete standards. Sprayed concrete is designed following the rules for cast concrete (i.e. ultimate limit state (ULS) and serviceability limit state (SLS) design cases), using material parameters confirmed by adequate testing. The construction methodology (sprayed or cast in situ) has no impact on the validity of the design rules.

However, the manufacturing process affects sprayed concrete in certain ways:

- Sprayed concrete is applied in layers, which may lead to the interface between layers having an impact on the mechanical behaviour of the overall concrete matrix. This potential for a change in behaviour compared with unlayered concrete should be considered in the specification of the concrete and the associated testing.
- Sprayed concrete mixes are mainly high-performance concrete mixes containing several admixtures including accelerators. The interaction between these admixtures should be verified in the mix development process. Mixes should be tested for the maximum and minimum envisaged accelerator content, and these limits should be observed on site. Certain combinations of admixtures and base materials can lead to long term strength loss and/or softening that should be avoided. The sprayed concrete or admixture supplier's advice should be sought and additional testing should be considered where necessary.
- The material parameters of sprayed concrete should be verified by cores taken in situ from the lining, and/or cores and beams cut from sprayed boxes.
- Sprayed concrete maturity-dependent performance requirements are relevant for design and construction. For construction, the target criterion is rapid strength gain. The designer may postulate strengths at certain concrete ages. Performance requirements should contain upper and lower limits.
- The site team should be involved as early as possible in the design and specification of sprayed concrete.
- The hierarchy of protection in the Management of Health and Safety at Work Regulations 1999 requires that if elimination of a hazard is not reasonably practicable then engineering means should be used to reduce risk. With reference to clause 7.7.3.3 of BS 6164:2019, it is required that, as sprayed concrete is a highly engineered material, mix design should include the reduction of dust emissions as a design parameter.

8.2. Mechanical parameters and behaviour

8.2.1 Sprayed concrete

Sprayed concrete behaves in a similar manner to traditionally used cast in situ concrete but the mix design and placement process result in different characteristics. In addition, material parameters depend on the orientation of the stresses and strains relative to the spraying direction.

The performance requirements for sprayed concrete emphasise different aspects of the concrete compared with in situ concrete, the most prominent being early age strength development. The resulting mix designs should be optimised to achieve associated performance criteria, and might exhibit atypical behaviour regarding standard parameters – for example, the elastic modulus and shrinkage development with age or maturity. It is recommended that performance requirements and associated testing for all mechanical parameters of sprayed concrete used in the design are specified considering age or maturity dependence.

8.2.2 Fibre reinforced concrete (FRC)

Steel or macro synthetic fibres can be included in sprayed concrete mixes to ensure ductile material behaviour of the cracked concrete, and to provide resistance in tension and flexure. Fibres have limited impact on the performance of un-cracked concrete.

FRC is a composite material with, ideally, isotropic material behaviour due to the random distribution of the fibres in the concrete. Both steel and macro synthetic fibres typically provide tensile resistance post cracking and ductility compared with plain concrete, by a combination of elasto-plastic fibre strain and, for steel fibres, the hook-ended fibres being gradually pulled out of the cracked concrete (de-bonding and then being progressively deformed as they are pulled out). The total strain of the concrete zone in which a fibre is anchored is, therefore, the sum of the strain in the fibre and the strain associated with the slip of the fibre.

This slip of fibres relative to the concrete is a key implicit performance criterion required for ductile material behaviour. A significant influence on this slip is associated with the bond strength of concrete, which increases with compressive strength of the concrete. Concrete strengths significantly above the assumptions made in the design can lead to increased slip resistance. This, in turn, can lead to stresses in the fibre exceeding its capacity, leading to fibre rupture and, in consequence, brittle behaviour, albeit at a relatively high strength. It is therefore necessary to select a fibre suitable for the actual strength of the concrete at the time the relevant ULS and SLS occur (minimum reinforcement). Rules ensuring sufficient ductility of FRC can be found, for example, in the fib Model Code 2010.

FRC can be classified by its behaviour in tension (direct tension and flexural tension). Behaviours include the following:

- Brittle failure. In concrete, where the fibre content is low, the fibres have no or very limited impact on the post crack behaviour of the concrete. The FRC design should follow plain concrete design rules.
- Semi-brittle. The fibres do not provide tensile performance fulfilling the ductility criteria of structural design codes, but a reproducible and relevant amount of ductility (e.g. the D1/D2/D3 classes from BS EN 14487-1:2022). While the resulting FRC cannot be classified as reinforced concrete, the increased amount of ductility can be implicitly considered in the design.

▨ Ductile-strain softening. FRC whose tensile resistance drops after the first crack is observed is classified as ductile-strain softening. Such an FRC can still fulfil the ductility requirements of the applied design codes and guidelines (e.g. fib Model Code 2010) and thus can be designed as structural reinforced concrete. For FRC mainly designed for ULS, this material behaviour is adequate.

▨ Ductile-strain hardening. Strain hardening behaviour is characterised by increasing tensile resistance after the point of the first crack. This performance would typically only be achieved by using high performance fibres and/or a large amount of more conventional fibres. Strain hardening sprayed FRC is typically not used in SCL tunnelling applications, except where the lining might go into direct tension. It should be noted that high fibre content as well as the high-performance fibres are likely to have a major impact on the ability to pump and spray the material. It is not recommended to specify ductile-strain hardening FRC unless the workmanship issues have been assessed with the supply chain.

The ductility modes of FRC specimens tested with a three-point flexural beam test (EN 14651) are indicated in Figure 8.1. The test load is converted into a corresponding residual flexural tensile strength as per BS EN 14651:2005, equation (4).

The actual performance of fibres in concrete depends on many variables – for example, fibre material, fibre shape, fibre content, cement content, aggregate grading, water/cement ratio, admixtures and so forth. Specifying these parameters, to achieve a reproducible performance of the FRC mix, is typically not possible due to the number and the interdependency of the parameters. Hence, FRC is generally specified in terms of performance criteria, which can include the following mechanical parameters:

▨ Minimum and maximum compressive strength at 28 days.
▨ Strength development over time (short term up to 28 days/long term beyond 28 days).
▨ Direct tensile strength.
▨ Flexural tensile strength (residual).

It is generally not recommended to specify fibre content or fibre type in the design phase as this removes some options for mix optimisation in the supply chain.

8.2.3   FRC mechanical parameters

The presence of fibres in the concrete matrix generally only influences the behaviour of the composite material in tension, and only in its cracked state.

8.2.3.1   Compressive strength and strain

Compared with unreinforced concrete, the behaviour of FRC in compression remains virtually unchanged for fibre dosages relevant to practical applications. The strength/strain relations from relevant concrete design codes – for example, BS EN 1992-1: Eurocode 2: Design of concrete structures – can be used without modifications.

8.2.3.2   Tensile strength

The spraying process does not influence the tensile behaviour of concrete. FRC offers post cracking residual tensile strength that can be assessed from the results of flexural tensile strength tests, tensile splitting tests, or by direct tensile testing.

Figure 8.1 Ductility modes of FRC

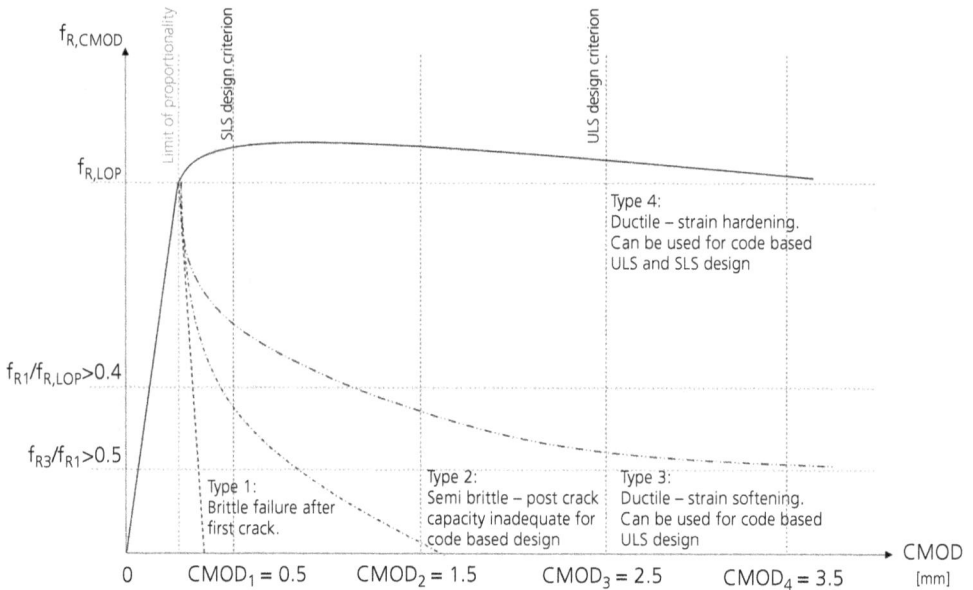

8.2.3.3 Flexural tensile strength

The flexural tensile strength is usually taken as the residual flexural tensile strength after the first crack (limit of proportionality (LOP)) for permanent and transient ULS and SLS. The flexural residual strength is generally specified in terms of flexural resistance of test beams at pre-defined strain levels associated with SLS and ULS (using the EN 14651 test f_{R1} and f_{R3} in fib Model Code 2010; and f_{R1} and f_{R4} in RILEM TC 162-TDF: Test and design methods for steel FRC).

8.2.3.4 Shear strength

It is commonly accepted that using fibres leads to a shear enhancement of concrete; however, the resulting effect is difficult to quantify. It is recommended that design rules for plain concrete are used for FRC in shear.

8.2.3.5 Strain limits

The compressive strain limits of FRC remain unchanged compared with conventional concrete. Strain limits in accordance with the design codes apply. The tensile strain of FRC is limited to 2.5% for ULS and SLS design in design codes such as the fib Model Code 2010.

FRC can be designed to sustain higher strains before failure, but the design of such is outside the scope of current structural design codes and standards, like the fib Model Code 2010 or the Eurocodes.

Further information on the characterisation and testing of FRC material parameters can be found in the ITAtech (2016) *Guidance for Precast Fibre Reinforced Concrete Segments, Volume 1: Design Aspects*.

8.2.4   Time dependency of sprayed concrete

The influence of time dependency on sprayed concrete design and construction are discussed in Chapters 11 and 12. The following aspects of time dependency should be considered:

▨ The immediate strength development between 1–24 h. Strength development over this time is crucial for the performance of the lining relative to immediate elastic stresses caused by relaxation of the ground. Apart from having a significant influence on the deformation of the tunnel lining, a minimum strength must be achieved to render areas safe to work in or to undertake the next advance.
▨ The strength development up to 28 days.
▨ The strength development post 28 days. While the concrete is specified in terms of its 28-day strength, it is important to understand the long-term strength gain and to verify that the material does not deteriorate over time. The material behaviour post-28 days and appropriate testing should be specified.

The accelerated hardening of sprayed concrete is achieved mainly by either using rapid hardening cements, or accelerating admixtures, or a combination of both.

The early age strength development is typically characterised by the curves presented in BS EN 14487-1:2022, as shown in Figure 8.2.

Figure 8.2 J curves of SCL strength development

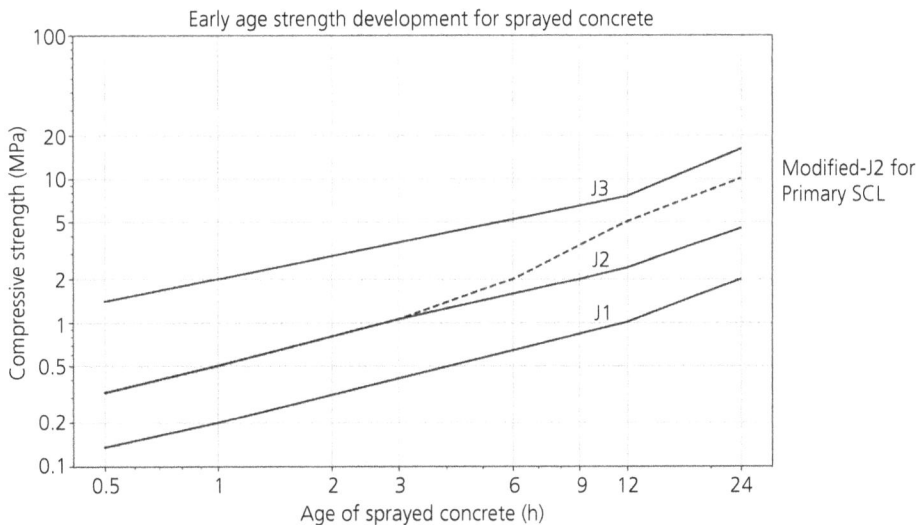

These curves can be used as a baseline for projects but should be reviewed and, if necessary, modified to suit project specific requirements. However, typically:

▨ J1 – is suitable for placing thin layers on a dry base without special load bearing requirements
▨ J2 – is suitable for placing thicker layers (including overhead) at a high delivery rate
▨ J3 – is used under special circumstances such as high ingress of water.

8.3. Designing for durability
8.3.1 Lining permeability

The quality of the sprayed concrete lining is directly related to its permeability. Specifying a maximum permeability and testing allows the assessment of durability against reinforcement corrosion.

A reinforced concrete lining that is designed to provide flexural resistance in the long-term ULS requires consideration of the water penetration and the associated corrosion mechanisms relevant to the reinforcement.

Performance requirements for this aspect of durability are not standardised in BS EN 206-1 or BS 8500 and assume the use of in situ concrete. There is no direct link in these codes between exposure classification and concrete permeability. The damage mechanism has been considered through the link between exposure classification and cover.

Conducting permeability tests delivers a criterion to compare the compaction and durability of sprayed concrete against the baseline underlying the durability performance requirements of BS EN 206 and BS 8500. Achieving a similar level of permeability indicates an equivalent level of compaction of the installed concrete.

The BTS *Specification for Tunnelling* contains guidance on minimum performance requirements for permanent sprayed concrete linings. Further detailed information can be taken from Concrete Society (2008) *Technical Report 31, Permeability Testing of Site Concrete*.

8.3.2 Reinforcement corrosion

When steel fibres are added to concrete, there is no cover to fibres at the surface. As the concrete carbonates, the steel fibres corrode, but due to their small volume there are insufficient expansive forces generated to disrupt the concrete, as would be the case with conventionally reinforced concrete. Since the fibres are small and electrically discontinuous, they are not subject to galvanic corrosion.

The ingress of chloride ions can depassivate the concrete and enable corrosion to progress. This depassivation process can lead to a possible loss of fibre performance. It is possible to assess the corrosion of the fibres and the subsequent failure of the FRC material in tension by modelling the progression of the carbonisation through the cross-section and assuming an appropriate corrosion rate for the fibres together with a failure criterion. This reduction in capacity should be considered in the design. Alternatively, the contribution of the fibres within a zone determined to suffer carbonation or chloride ingress over the design life of the structure can be ignored.

Various tests undertaken to date indicate that FRC shows roughly comparable resistance to crack-induced corrosion as conventional bar reinforced concrete. A summary of applicable tests is contained in ACI 544.5R-10: *Report on the Physical Properties and Durability of Fibre-Reinforced Concrete*.

Given the limited contribution of the fibres to the long-term performance of the tunnel lining it would be reasonable to use Table 7.1N of BS EN 1992-1: Eurocode 2 (for conventionally reinforced members) to determine crack width limits.

8.3.3 Drying shrinkage

Shrinkage can be assessed by testing, advanced material modelling, and using design guidance such as CIRIA (2018) *C766: Control of Cracking Caused by Restrained Deformation in Concrete*. Minimum requirements regarding the drying shrinkage of coarse aggregates are provided in the BTS *Specification for Tunnelling*. Shrinkage strain calculations should always be based on the actual expected strength of the concrete where there is a significant difference to the specified value.

Shrinkage strains and strains resulting from loads, and the average distance between cracks, are the governing parameters required for assessing the average crack width of concrete. Shrinkage requirements are typically only relevant for permanent sprayed concrete linings.

Only strain hardening sprayed concrete – that is, concrete with bar reinforcement and/or with a substantial amount of high-performance fibres – can be designed to comply with fixed crack requirements using design codes. If crack limits are to be observed using strain softening FRC, the material behaviour for this design check should be validated by appropriate testing.

8.3.4 Ductility – maximum concrete strength

Bar reinforced concrete exhibits ductile behaviour. The minimum level of ductility for reinforced concrete structures is typically defined by a minimum reinforcement quantity, which in turn is designed for the tensile stresses of the concrete immediately before the first crack.

The minimum reinforcement check is an SLS check and thus based on mean material parameters. It is important to consider the actual expected concrete strength immediately before the first crack not the specified value, which can be considerably lower. Otherwise, the reinforcement might be insufficient to accommodate the tensile stresses transferred from the concrete matrix, leading to excessive strains of the reinforcement which, in case of fibres, might result in fibre failure.

The fib Model Code 2010 (5.6.3) states the following minimum requirements for fibre reinforcement concrete sections:

- The flexural resistance of the FRC at the strain level used for SLS design (at CMOD = 0.5 mm, f_{R1k}) must be greater than 0.4 times the flexural resistance at LOP (first crack, f_{Lk}).
- The flexural resistance of the FRC at the strain level used for ULS design (at CMOD = 2.5 mm, f_{R3k}) must be greater than 0.5 times the flexural resistance at the strain level used for SLS design (at CMOD = 0.5 mm, f_{R1k}).

CMOD is the crack mouth opening displacement given in mm, defined in BS EN 14651:2005.

For this check, the designer should assess the maximum concrete strength at which the first crack can occur and obtain confirmation from the fibre manufacturer that their fibre is suitable for concrete of this strength.

emerald PUBLISHING

ice

British Tunnelling Society
ISBN 978-1-83608-693-2
https://doi.org/10.1108/978-1-83608-690-120251010

Chapter 9
Designing for water

9.1.　Water management strategy

The decision to keep external groundwater out of the tunnel on a permanent basis or to allow some passage of water through into an internal drainage or water management system is based on a range of factors, including the following:

- The potential impact of water ingress on the internal contents of the tunnel.
- The balance of costs between designing to take the full water loading onto a lining or the longer-term costs of pumping to remove water or maintaining drainage systems.
- The potential impact of allowing long-term drainage on the groundwater profile of the area the tunnel passes through, including effects on local water extraction and the long-term settlement of buildings and services.
- The risk of failure of the chosen system and the costs to rehabilitate/repair.

The costs of installing and maintaining a drainage system should be considered on a whole life basis. Systems using pumps would need to allow for regular maintenance, testing and replacement costs as well as the electrical input. Where keeping the water out is safety critical, back-up systems may also be required. While gravity drainage would avoid the majority of these costs, this option may be limited by tunnel alignment, risk of contamination of the external water table through water passing through the tunnel environment and storage capacity before treatment. Drainage systems may become clogged, particularly through leachates in the groundwater. This may result in the capacity of the lining being exceeded as water pressures are generated behind the lining which the lining has not been designed for, or the system backing up and water collecting in the tunnel.

Allowing long-term drainage of soft ground can lead to consolidation and long-term settlement, sometimes in excess of that associated with the initial construction of the tunnel. The potential damage to overlying structures and services would almost always make that unacceptable and efforts to minimise changes to the groundwater regime should be part of the design criteria. Where water is being extracted, the permanent drawdown in the water table may not be environmentally acceptable and may upset the balance of water available at other locations or result in contamination.

It might, therefore, seem that a fully tanked (where a complete waterproof barrier has been applied to the exterior to prevent water from entering system) would always be preferable as it avoids the risks and costs outlined above. However, designing for the water loads will likely mean thicker, more heavily reinforced and, therefore, more costly linings. In addition, if the waterproofing is compromised over the tunnel's lifetime, then additional water management, not envisaged in the design, may need to be installed. While there are costs associated with the maintenance and repair of a drainage system, these costs are known and can be planned for and may be quickly overshadowed by the unexpected costs of repairing a fully tanked system, if it is practically possible at all.

The amount of water inflow that is considered acceptable within the tunnel structure will often be included in the client's specification. In many industries, there are specified classes of water-tightness that are commonly used as a reference (e.g. rail industry standards where water might damage track or signalling and raise unacceptable risk to passengers). In some cases, water ingress is acceptable in certain areas (e.g. below tunnel axis) but should not be visible in others (in public areas, mechanical and electrical control rooms etc.). For tunnels constructed to transfer water, including sewage, the risks associated with water escaping out of the tunnel would often necessitate a fully watertight solution. Refer to the BTS *Specification for Tunnelling* for typical values.

9.2.   Designing for fully tanked solutions

When designing a fully tanked solution, there are a range of design solutions available. These include the following:

- Sheet waterproofing membranes (factory produced materials that should be installed within the tunnel in rolls and welded at joints between adjacent rolls).
- Spray applied waterproofing membranes.
- Designing the structure of the tunnel to prevent water ingress using watertight concrete.

Each of these design solutions has benefits and disadvantages depending on the overall design solution for the tunnel. The balance of risk, both to cost and programme, as well as health and safety, should be assessed against the project requirements.

9.3.   Sheet waterproofing membranes

Sheet waterproofing membranes have the longest track record for use in tunnels. Traditionally installed over a drainage fleece inside a primary lining, sheet membranes are supplied in rolls of material that can be heat welded together to form a continuous barrier. Welds can be tested and linings divided into compartments using waterbars so that any subsequent water ingress can be contained and easily targeted with grouting systems. More recent innovations in sheet membranes include double sheeting, where the voids in between can be injected with flexible acrylic, and membranes that can be adhered directly to the substrate.

Sheet membranes are typically installed by specialist supply teams using bespoke gantry systems for work at height, which allows access to the tunnel crown and walls. Water ingress through the substrate can be temporarily or permanently channelled away to allow installation of the membrane. Sheet membranes can be installed in collaboration with a drainage system, and, in drained lining designs, the geotextile fleece and sheet membrane are terminated to a pipe, normally installed at either the haunch or base of the invert. For undrained systems the fleece and membrane extend around the full perimeter of the tunnel.

The materials are flexible and can allow for large deformations or tensile forces. Membranes should be specified with high fire resistance (BS 6164:2019: clause 7.7.3.4) and can be specified with high thermal and chemical resistance. There are historic cases of damage to membranes through plasticiser loss (within the membrane) in locations of high water flow and, therefore, where high flows are anticipated, further consideration may be needed. In cases where the tunnel has a uniform cross-section and a cast in situ secondary lining, sheet waterproofing membrane would often be the most cost-efficient waterproofing solution available.

9.4.   Spray applied waterproofing membranes

Spray applied waterproofing membranes are designed to bond to their substrate, reducing the risk of water ingress by eliminating water paths along the interface. Most spray applied membranes provide a bond to the secondary concrete lining, and therefore, if water does penetrate the membrane, the water paths along the interface with the secondary lining are limited by this bond. These bonds are dependent on the condition of the substrate and degree of water ingress through the primary sprayed concrete lining. This, in turn, can necessitate smoothing processes to the primary lining and treatment of active water ingress through installation of drainage conduits or pre-treatment by chemical injection.

Further design guidance on spray applied waterproofing membranes, including typical materials and performance can be found in the ITAtech (2013) *Design Guidance for Spray Applied Waterproofing Membranes*. This guidance document has been written to assist tunnel designers, contractors and owners in understanding the benefits of and limitations in the use of spray applied waterproofing membranes in excavated tunnels and shafts, and to provide guidance in developing specifications and design details. The hazardous nature of spray applied waterproofing materials should be considered an essential part of the material selection process as required by BS 6164:2019, clause 16.10.

As with sheet membranes, the application of spray applied membranes is typically carried out by manufacturer approved applicators; however, some systems can be applied by the main works contractors as long as they have received training and approval from the supplier. The performance of spray applied membranes is dependent on achieving good coverage and the membrane having full bonding and crack bridging properties. The advantage of spray applied systems is that the shape of the structure has very little impact on the application and so complex geometries and junctions can be more easily addressed. They are also considered better at facilitating the application of a sprayed concrete secondary lining.

9.5.   Watertight concrete

Cast in situ concrete can produce a low permeability material and can be reinforced to provide crack control suitable for a water retaining structure. The addition of a waterproofing layer outside a cast secondary lining could therefore be considered as unnecessary, but careful detailing of joints is essential. For the main part, the design of watertight concrete would focus on the control of cracking through reinforcement and curing control measures. Alternatively, polymer-enhanced concrete is an option. However, its use in the UK tunnelling industry is not as well established.

While a sprayed concrete mix can be demonstrated to have very low permeability in samples – due to the nature of the spraying process and the joints created between advances – it would need to be proven to be an effective solution if not combined with additional measures. The sequencing and control of joints would need to be carefully specified.

9.6.   Junctions between structures

Independent of the chosen waterproofing system, care should be taken when designing waterproofing at junctions between structures and at changes of section. These are often locations where the relative stiffness of the adjoining sections may lead to differential movements; increasing the likelihood of cracking that would allow water ingress. Design solutions to connect these structures locally using reinforcement may transfer the location of cracking away from the junction, but, in most cases, differing relative movements may result in damage to the lining and, therefore, additional waterproofing measures should be considered in these areas.

In locations where movements are likely to occur over the life of a structure, it is suggested that redundancy and repeatability are built in. Hydrophilic materials can offer limited repeatable shrink/swell behaviour and have the capacity to adjust to their changing environment. However, the maximum number of wet/dry cycles should be obtained from the supplier and assessed against the corresponding design criteria. Re-injectable injection hoses, if correctly installed and injected, may allow cracks to be resealed with minimal impact. Locations of the termination boxes should be designed to be easily accessible over the life of the tunnel.

9.7.   Impacts on design/construction

The choice of waterproofing system impacts the design of the tunnel lining. In many cases, the decision cannot be taken in isolation but alongside consideration of geology and secondary lining methodology. Where a cast in situ secondary lining is chosen – for example, for long tunnels of uniform section – both sprayed and sheet membranes can be used. In tunnels where a sprayed secondary lining is used, a spray applied membrane would be preferable as applying sprayed concrete to sheet membranes has not been embraced by UK clients.

In many cases, a cast in situ concrete secondary lining will have a lower carbon footprint than the same thickness of sprayed concrete due to lower cement content, fewer admixtures, smaller geometrical tolerances and less wastage.

In all cases, the potential for water travelling through the waterproofing system substrate during construction of the tunnel should be considered. In soft ground tunnelling, water ingress is either naturally low – for example, due to the low permeability of clay strata – or the water table should be locally controlled during construction to allow safe installation of the primary support. Some depressurisation (see BS 6164:2019: clause 9.1.2) or water management techniques can be maintained during the installation of the waterproofing system if the substrate needs to be dry. Local treatment using grouts, and so forth, can seal water off until a spray applied membrane has been installed and is cured. However, the balance between these requirements may make a sheet membrane more practicable.

British Tunnelling Society
ISBN 978-1-83608-693-2
https://doi.org/10.1108/978-1-83608-690-120251012

Chapter 10
Analytical solutions

10.1.　Introduction

Although numerical methods of analysis have become standard practice for SCL design as they can capture complex geometries with realistic soil behaviour and construction sequencing, analytical (sometimes called closed-form) solutions can be used for relatively simple geometries and ground conditions. They are described in the BTS (2004) *Tunnel Lining Design Guide*.

The advantages of analytical solutions are that they can be solved by hand or in simple spreadsheets. This enables rapid answers to be found, and sensitivity analyses may be performed quickly by varying input parameters. The simplicity allows easier interpretation, and the understanding gained is useful when introducing more complexity using numerical models.

Analytical solutions may also be used to validate numerical modelling programs by simulating the same boundary conditions and using the same constitutive models.

This principle of gradually increasing complexity by beginning with simple models, moving on to analytical solutions, then simple numerical models, then incrementally increasing complexity of geometry and constitutive models, is a good approach. At every stage, the effect of the new element of complexity in the model can be understood, giving confidence to move on. The problem with jumping straight into a sophisticated model is that the effect of the different layers of complexity or their relative importance may not be understood. In addition, if the results are not sensible, debugging is difficult, whereas with the incremental approach it is usually the case that it is the latest change or addition that is the cause of the problem.

Analytical methods are simple and inexpensive to undertake, and it is recommended that these methods are exploited at the earliest stages of tunnel and shaft SCL design.

However, analytical solutions have their limitations. They are limited in the complexity of geometry that can be modelled, the ability to model different strata, the variability of geotechnical parameters with depth, non-hydrostatic groundwater conditions, construction sequences, construction loads, and the complexity of the geotechnical and sprayed concrete constitutive models.

Analytical solutions are generally two-dimensional (2D) idealisations that assume the ground is a homogeneous continuum and the tunnel is circular. These solutions do not explicitly address interaction between adjacent tunnels/excavations, requiring the use of superposition or rules of thumb to address the additional stresses induced in the lining due to adjacent construction.

There are a variety of analytical solutions available that can be used for designing aspects of SCL tunnels and shafts and include:

- Circular tunnels or shafts
- Face stability
- Junction design
- Ground movements.

10.2. Circular tunnels and shafts

10.2.1 In situ ground stress

The simplest model of a tunnel or shaft is to directly apply the in situ ground stress for the relevant design situation to the extrados of the lining. This ignores the construction sequence, ignores soil-structure interaction, and does not allow the calculation of bending moments in the lining. However, it provides an upper bound to the axial (hoop) force in the lining which can be used for initial sizing of the sprayed concrete shell, subject to adequate allowances for deformation and build tolerance being made.

10.2.2 Continuum analytical models

Continuum analytical models include the effect of soil-structure interaction and unequal ground stresses. A number of closed-form models have been developed and generally assume that the ground is an infinite elastic, homogeneous, isotropic medium. The interaction between ground and a circular, thin-walled lining is assumed to occur under 2D plane strain conditions. Essentially, the tunnel lining is in a stress field with a vertical and horizontal component of ground stress. The ground and the lining are assumed to be linear elastic, defined by a Young's Modulus and Poisson's ratio. The interface between the lining and the ground can allow either full-slip or limited or no slip conditions along the ground lining interface, which represent the two limiting cases with the true situation lying somewhere in between. The approach is described in the following key papers:

- Muir Wood (1975) The circular tunnel in elastic ground. Also, Curtis (1976) A discussion note [on Muir Wood (1975)]. Together, they are referred to as the Curtis/Muir Wood Equations. The equations were developed to include viscoelastic ground and included radial deformation due to tangential stress.
- Einstein and Schwartz (1979) Simplified analysis for tunnel supports. This developed different alternative expressions for flexibility and compressibility ratios.
- Duddeck and Erdmann (1985) *Structural Design Models for Tunnels*. This suggested that reduction factors to the full applied ground stress could be considered.

When applying closed-form solutions of this type to the design of a tunnel lining, the decision about what value of horizontal and vertical stress to apply is the governing influence on the thickness of the lining and/or the amount of reinforcement required. The 3D nature of a tunnel face and the redistribution of ground stresses that occur prior to installation of the lining mean that applying the pre-construction in situ stress state, particularly in strongly over-consolidated soils, would be unrealistic. A judgement needs to be made about what stresses should be applied to the tunnel lining.

Results from finite element analysis have indicated that these types of solutions should only be applied to tunnels where the depth to tunnel axis is greater than 1.5–2 times the external tunnel diameter (Duddeck and Erdman, 1981; Vu and Broere, 2018; Vu *et al.*, 2017).

10.2.3 Convergence-confinement method

The convergence-confinement method can be used in a wide range of ground conditions to predict ground movement and tunnel support measures. The method is described in the following key papers:

- Hoek and Brown (1980) Underground Excavations in Rock.
- Panet and Guenot (1982) Analysis of convergence behind the face of a tunnel.
- Eisenstein and Branco (1991) Convergence–confinement method in shallow tunnels.
- AFTES (2001) Recommendations on The Convergence-Confinement Method.

This method has developed over time and now includes the effects of plasticity according to the Mohr–Coulomb or Hoek–Brown yield criteria, creep in the ground, gravity effects, the timing of support by way of the geometric delay parameter, support type (sprayed concrete, concrete, steel sets and rock-bolts) and mined or TBM-driven tunnels.

The method is valid only for circular cross-sections where K_0 (coefficient of earth pressure at rest) is close to 1.0 and the tunnel is constructed using full-face excavation techniques. In addition, no information is given on the distribution of bending moments and shear forces in the lining. For more complex applications of the method, a numerical solution is required.

10.3. Face stability

The stability of tunnel faces can be assessed for upper and lower bound conditions using stability ratios. A detailed description of all the analysis methods, with example calculations, can be found in Jones (2022). Design guidance may be found in DAUB (2016) 'Recommendations for face support calculations for shield tunnelling in soft ground'. A review of calculation methods is expected to be produced by ITA Working Group 2 in the next year.

The most common methods of analysis are described in the following papers:

- Broms & Bennermark (1967) Stability of clay at vertical openings.
- Kimura and Mair (1981) Centrifugal testing of model tunnels in soft clay.
- Davis *et al.* (1980) The stability of shallow tunnels and underground openings in cohesive material.
- Perazzelli and Anagnostou (2017) Analysis method and design charts for bolt reinforcement of the tunnel face in purely cohesive soils.
- Pferdekämper and Anagnostou (2022) Undrained trapdoor and tunnel face stability revisited.

The above papers consider stability in undrained soils. The papers below consider solutions for tunnels in drained soils using Mohr-Coulomb criteria (c' and ϕ'):

- Leca and Dormieux (1990) Upper and lower bound solutions for the face stability of shallow circular tunnels in frictional material.
- Anagnostou and Kovári (1996a) Face stability conditions with earth-pressure-balanced shields.
- Anagnostou and Kovári (1996b) Face stability in slurry and EPB shield tunnelling.
- Anagnostou G (2012) The contribution of horizontal arching to tunnel face stability.

10.4. Junction design

The design of SCL junctions is a 3D problem that can only be roughly approximated by closed-form solutions. Available analytical methods are mainly derived from a 'hole in a plate' approach, in which the plate is 2D and plane stress. In consequence, the impact of stresses acting normal to the surface of the plate or the effect of curvature of the plate are not considered. These effects may be minor for openings that are small relative to the size of the parent tunnel but can have a major impact on relatively large openings.

Analytical methods are not considered best practice for the design of tunnel junctions because they do not include the construction sequence, soil-structure interaction or the 3D geometry. They will not predict any bending moments, which can be very significant. It is recommended 3D structural and/or geotechnical models are used as this allows assessment of the construction sequence, the 3D effects and the soil-structure interaction.

The opening design can be carried out on a structural (bedded plate or similar) 3D numerical model calibrated to yield similar hoop and shear forces as well as bending moment results as a continuum numerical model in its unopened stage. The opening is then introduced in the calibrated structural model, resulting in an upper bound approximation of the stresses around the opening. This will not model soil-structure interaction properly because the ground is modelled by independent springs. Thus, the ground pressures at the opening will not redistribute by way of ground arching to the lining around the opening.

A better approach is to use a 3D continuum model that includes all the construction stages for the parent tunnel, the creation of the opening and the construction of the child tunnel. This ensures that soil-structure interaction, the construction sequence and the 3D geometry of the tunnels are included explicitly.

The designer should consider that the 3D situation at junctions and openings results in much increased complexity not only of the design, but also of the construction. It is recommended that the design is developed to accommodate increased construction tolerances and local imperfections.

10.5. Ground movements

Tunnel and shaft construction can induce ground movements that could result in deformation of surrounding structures and infrastructure. The prediction of tunnel induced ground movements and the resulting structure deformation is a key issue in the planning of SCL tunnel projects because, by its very nature, open face SCL tunnelling will likely induce higher ground movements than closed face mechanical tunnelling methods.

The prediction and effects of ground movements are discussed in the BTS (2004) *Tunnel Lining Design Guide* (2004: Chapter 7) and in Jones (2022: Chapters 1, 12, 13) *Soft Ground Tunnel Design*. Detailed design approaches for predictions of ground movement and damage assessment are beyond the scope of this document.

The empirical methods commonly used for predicting ground movements are described in the following publications:

- Mair *et al.* (1996) Prediction of ground movements and assessment of risk of building damage due to bored tunnelling.
- Mair *et al.* (1993) Subsurface settlement profiles above tunnels in clays.

- New and O Reilly (1991) Tunnelling induced ground movements: predicting their magnitude and effect.
- New and Bowers (1994) Ground movement model validation at the Heathrow Express Trial Tunnel.
- O'Reilly and New (1982) Settlements above tunnels in the UK, their magnitude and prediction.
- Attewell and Woodman (1982) Predicting the dynamics of ground settlement and its derivatives caused by tunnelling in soil.

A review of shaft settlement prediction methods may be found in the following:

- New (2017) *Settlements due to Shaft Construction.*
- Jones (2022) *Soft Ground Tunnel Design.*
- Taborda *et al.* (2024) A methodology for improved predictions of surface ground movements around shafts.

Excavation face stability and rapid closure of the ring are critical in controlling ground movements when using SCL techniques. Toolbox items such as ground treatment, dewatering, face nails and spiling can control face stability in challenging ground conditions. Careful sequencing to avoid unsupported faces, using pocket excavations and reducing advance lengths can also be used to improve stability and hence reduce ground movements.

Volume losses on most SCL projects in soft ground in the UK are typically less than 1.5% and the assumed volume loss for SCL design is therefore commonly specified as 1.5%. The applicability of this value should be checked for cases of particularly large tunnels where ring closure might be delayed or where cover is low.

Actual volume loss in SCL tunnels from case histories is shown in Table 10.1. Note that in certain cases the use of compensation grouting masks the effect of settlement, making it significantly more difficult to make accurate estimates of actual volume losses for those projects.

Table 10.1 Precedent volume loss information

Project and SCL Works	Predominant ground conditions	Volume loss measured at end of construction
Jubilee Line Extension Project, Waterloo Station	London Clay	1.0% – 1.5%
Heathrow Express trial tunnel	London Clay	1.1% – 1.4%
Heathrow Express main works – T4 Station platform tunnels	London Clay	0.6% – 0.9%
Heathrow Express baggage transfer tunnel	London Clay	0.5%
Kings Cross Station Redevelopment Phase II	London Clay	0.7% – 1.3%
Crossrail – Whitechapel Station platform	London Clay	1.0% – 1.3%
Crossrail – Liverpool St Station platform	London Clay	0.9% – 1.4%
Crossrail – Stepney Green caverns	London Clay	1.3%
Hindhead tunnel	Sandstone	0.5% – 1.0%
Round Hill tunnel (portal areas)	Chalk	0.1% – 0.5%

For clay soils, short-term volume loss can be estimated using the 'load factor approach'. The 'load factor' is the inverse of the factor of safety on stability, and therefore this empirical approach relates stability to the magnitude of ground movements. This relationship was first developed based on centrifuge tests by Mair *et al.* (1981), and a meta-analysis of case histories was performed by Macklin (1999). This was later slightly refined by Dimmock and Mair (2007). In practice, this means that volume loss is dependent not just on soil behaviour, but also the size of the exposed face, the depth of the tunnel, any surface surcharge, the unsupported length of the heading, and the effectiveness of any mitigation measures used to improve stability.

emerald
PUBLISHING

ice
Publishing

British Tunnelling Society
ISBN 978-1-83608-693-2
https://doi.org/10.1108/978-1-83608-690-120251013

Chapter 11
Numerical modelling

11.1. Introduction

Numerical modelling methods have become standard practice for tunnel design as they can capture complex geometries and realistic soil response mechanisms and behaviour as well as accounting for interaction between structures. An appropriate numerical model of an underground structure can give a good and communicable description of the structural behaviour, indicate risks and highlight issues that deserve additional attention during design or construction, thus providing substantial aid to the project development.

Numerical modelling aims to capture the complex behaviour introduced by soil-structure interaction, lining forces and deformations. It allows assessment of face stability, damage to adjacent tunnels or other underground structures, decisions about the required monitoring and prediction of the ground behaviour.

A Good Practice guideline has been published by CIRIA (2020: Report C791) regarding management of numerical modelling in a broad field of geotechnical engineering that provides a framework for those who oversee the numerical modelling process, undertaking the analyses and use the results to get a successful outcome.

11.1.1 Limitations

Based on certain approximations and simplifying assumptions, models are used to simulate actual phenomena, such as the soil-structure interaction. Consequently, the efficiency of a numerical model relates to the amount and purpose of the information available and extracted from the analysis. Additionally, the modelling effort should be proportionate to the risk to be mitigated to ALARP and also potential benefit in terms of design efficiency and, therefore, saving in material, cost and programme.

As more sophisticated models have become technically feasible, the modelling worktime has shifted from the computational time to the time needed for building, troubleshooting and post-processing the model. The designer has considerable freedom to decide which modelling approach is to be selected for different problems. Standard approaches in tunnelling are to use two-dimensional (2D) plane strain or axisymmetric models, 3D wished-in-place approaches, 3D-staged approaches (simulating the excavation advance steps), or a combination of these for different parts of the tunnelling project.

11.2. Model set-up

A typical procedure for building a tunnel numerical model comprises seven main steps:

(a) Pre-modelling decisions. An investigation of the best modelling approach to determine whether two dimensional (plane strain or axisymmetric), 3D (wished-in-place or staged) or a combination of these using linear or non-linear models is appropriate. Other key items decided at the inception of the modelling procedure include the geometry and level of geometric precision, parts of the structure in focus and the adjacent structures for inclusion in the model, as well as the size of the model and the boundary distances.

(b) Model dimensions. The ground block (the solid block that contains the tunnel structures) in the model should be large enough so that ground movement induced by tunnelling or any type of loading within the model should not be felt at the boundaries or be practically negligible (e.g. less than one millimetre other than at ground surface). As a rule of thumb and based on past research results (e.g. Möller, 2006; Potts et al., 2002), the minimum dimensions of the finite element mesh are to be chosen so that the horizontal length of the mesh from tunnel axis to the edge corresponds to four to five times the tunnel diameter, and the distance from the tunnel axis to the bottom of the model corresponds to about two times the tunnel diameter.

(c) Ground modelling. An engineered selection of the geotechnical input parameters for the model has to be consistent with any input data from the field and previously acquired knowledge (e.g. literature or previous projects in the area). This stage might require the re-assessing of factual data to facilitate the particular soil model used – for example, specific small strain or strain-hardening parameters, considerations for the groundwater seepage effects on soil properties, or ground improvement techniques. Specific aspects of the geotechnical setting such as anisotropy, heterogeneity, faulting and fissures should be considered in the selection of input parameters, and related uncertainties have to be managed.

(d) Modelling of structural components. Assumptions should be made as to what is relevant in the case of support components in the model (e.g. should lattice girders or face support be considered) and what assumptions should be made for the design (e.g. preliminary dimensions, material properties of the lining such as concrete strength and early age strength development, cracked/uncracked lining model).

(e) Excavation and support sequence. The excavation approach should be simulated in the model since it affects the soil-structure interactions and the requested results. As the development of the optimum shape and the excavation sequence are heavily interlinked, the definition of the support sequence is likely to be iterative.

(f) Meshing (and boundary conditions). These define the elements size/type, together with the discretisation at areas of interest and where stress concentrations are expected.

(g) Trial analysis/calibration. In order to identify faults and particularities in the model it is necessary to validate the approach by preferably running a simplified model (e.g. linear elastic analysis, with coarse mesh, and simple elements). Only when the behaviour of this simplified model is transparent and explicable can higher degrees of complexity be introduced, and the final optimised model calibrated. Calibration may also be achieved by re-modelling a previous project where monitoring data is available. A trial/calibration analysis should also investigate the effect of varying parameters (sensitivity/parametric analysis), such as the geometry of the model, meshing/discretisation and material parameters.

11.3.   Constitutive model

The selection of the soil constitutive model depends on the objectives of the analysis. Mohr-Coulomb and Hoek Brown are well suited failure criteria for soil and rock, respectively. Nonetheless, the Mohr-Coulomb and Hoek Brown constitutive models considered as linear elastic-perfectly plastic models may not be able to accommodate various aspects of ground behaviour in loading and drainage conditions. The user should be aware of the limitations of constitutive models (Carter and Liu, 2005).

Soils and rocks, as natural materials, are characterised by high irregularity – that is, discontinuities, inhomogeneity, anisotropy and non-elasticity. Moreover, groundwater induces higher degrees of complexity. Depending on the degree to which these are present, and the limitations in modelling and computational effort, it may be reasonable to ignore some or all of them in the substrate material.

An advanced numerical model capable of considering anisotropy with small strain stiffness showed very slight discrepancy in ground movements with respect to homogeneous assumption in a boundary value problem (Schädlich and Schweiger, 2013). While this indicates that the influence of anisotropy is limited for modelling soil, anisotropy needs to be considered to capture jointed rock mass behaviour.

Material nonlinearity in the model may be imposed by well-established constitutive laws – for example, from simple to advanced formulations, the Mohr-Coulomb, the Hoek-Brown, the Modified Cam-Clay (Muir Wood, 1990), the Hardening Soil (Schanz et al., 1999) and Hardening Small Strain Stiffness (Benz, 2006) models.

The simplest constitutive soil/rock model is continuous, homogeneous, isotropic and linear elastic. Such simple constitutive models can be appropriate for model areas where only elastic deformations are expected and are mainly used to reduce computation time for models with complex geometry.

For strata where a nonlinear constitutive model is required, the Mohr-Coulomb model can be used as a basic representation of plasticity in the realm of geotechnics. Beyond this model's weakness to describe certain critical state phenomena, its benefits are that the model parameters are easily derived, and the results yielded by a Mohr-Coulomb model are easily interpreted and implemented. The criterion assumes that failure occurs when the shear stress at any point in a material reaches a value that is linearly dependent on the normal stress in the same plane.

The Mohr-Coulomb material model (with pre-failure linear elastic stress-strain behaviour) is appropriate in drained conditions and undrained conditions, either as an effective stress analysis using drained parameters c', ϕ', E' and ν' or as a total stress analysis using undrained parameters c_u, E_u and ν_u. Generally, the Mohr-Coulomb model is considered appropriate for ULS-dominated design. More advanced models should be used if ground movement and pore water pressure output is required.

Advanced constitutive models including the hardening soil model and the hardening soil model with small strain stiffness can consider more complex aspects of soft ground behaviour. A coupled undrained analysis is possible with these models, removing the deficiencies of the Mohr-Coulomb models in effective stress analysis under undrained condition. These models require more input parameters than the Mohr-Coulomb model, but all can be obtained from standard soil mechanics tests. The following aspects of soft ground behaviour can be considered using advanced constitutive models:

- Stress dependent stiffness
- Plastic straining from the onset of initial loading that can be partially irreversible
- Difference between unloading and reloading stiffness
- Accommodating soil behaviour in the range of small strains
- Model dilatancy of soil under undrained condition in an effective stress analysis that explicitly considers pore water pressure.

Advanced models are particularly useful where tunnelling induced deformations can have a considerable impact on adjacent structures, such as existing tunnels and foundations. Where such effects are relevant, the use of advanced models represents best practice.

11.4. Short term versus long term

The long-term effects on the lining system of tunnels in London Clay have been highlighted by many researchers (e.g. Barratt *et al.*, 1994; Jones and Grand, 2024; Jones *et al.*, 2023; Mair, 2008; Mair and Taylor, 1997; Wongsaroj *et al.*, 2007, 2013). Increases in the lining forces and time dependent settlements have been observed following tunnel construction in many cases. This is an effect of pore water pressure redistribution over time and should be considered in the design of tunnel linings. The effect is illustrated in the Transport for London Guidance Document G055 (Figures 1–4). These or similar curves can be used to check the output of numerical models or to design with closed-form equations.

Short-term loads are the effects of excavation and stress re-distribution from the tunnel face onto the primary sprayed concrete lining support, taking into account the pore water pressure alterations (e.g. negative pore pressure in cohesive and fine-grained soil layers such as London Clay) induced in the ground by the excavation. The long-term reinstatement of stresses around the tunnels depends on the final pore water pressure influenced by the water proofing system, and the associated stress changes on the lining need to be accounted for by the tunnel permanent support system.

A simplified approach can assume a simulation of the tunnel construction and installation of the primary lining with undrained soil parameters (short-term situation), followed by the installation of the secondary lining. In this instance, the soil parameters are switched from undrained to drained and the submerged soil weight is applied in combination with the associated water pressures. This simplified approach allows for a conservative estimation of the lining forces and the surface settlements in the long term.

These approaches can be further simplified by designing the primary and secondary lining as individual structures. Such simplifications however often require conservative assumptions, in particular with regard to the magnitude of the loads, and allocation to the primary and secondary lining, respectively.

11.5. Lining properties
11.5.1 Overview

Primary sprayed linings are constructed using a process of sequential excavation and support, and the loading history determines the stress situation in the lining. The effects of this process should be considered in the modelling approach. An important aspect is the simulation of the concrete hardening (stiffness and strength gain) as the excavation advances. This interface is preferably captured in the design by associating minimum concrete strengths to construction stages.

To assess this effect two simple approaches are normally used, namely:

(a) The assignment of an estimated concrete stiffness in every modelled step, or
(b) The assignment of an averaged hypothetical modulus of elasticity.

In case *(a)*, the concrete strength gain and an age-dependent stiffness value are assigned for each concrete section installed behind the excavation face; the correct stiffness value can be extracted from the project specifications or established by stiffness-time dependence formulas found in codes and other literature (e.g. fib Model Code for Concrete Structures 2010). The performance requirements for strength gain of the concrete mix in the specification should reflect the minimum values identified in the design.

For case *(b)*, the stiffness value assigned to the lining has to capture an averaged soil-structure interaction over the main construction subsequence in which the concrete is yet to reach its long term design strength. For instance, an averaged 'green' Young's Modulus of sprayed concrete may be applied to the unclosed top-heading and bench/invert while Young's Modulus of the hardened concrete is assigned to the closed rings. This approach is useful for initial assessments but should not be used for complex sequencing. Guidance to assess the appropriate stiffness and indicative values may be found in the literature – for example, in Karakus (2007), John and Mattle (2003) and Pöttler (1990).

The aforementioned linear elastic approach with a stepwise increase of the Young's modulus in subsequent excavation stages may give realistic lining deformation, however, lining stresses are usually too high – in particular, if the lining is subjected to significant bending. Non-linear analyses are helpful to assess the behaviour of cracked concrete liners in the context of soil structure interaction and allow for a more realistic assessment of the stress redistributions in the linings after cracking.

An advanced constitutive material model has been developed for sprayed concrete by Schädlich and Schweiger (2014), which can account for the most important aspects of its behaviour; simulating the time-dependent strength and stiffness of concrete, strain hardening-softening in compression and tension as well as creep and shrinkage using Mohr-Coulomb failure criterion. It is evident that extra model parameters to be derived from lab tests (e.g. uniaxial multistage creep tests in compression and tension) to consider the time-dependent strength and stiffness of concrete.

Generally, concrete stiffness should be assigned an upper bound value when the aim of the analyses is the design against section forces. A lower bound value should be assigned when using the deformation output of the model for assessing ground movements, impacts on adjacent structures and derivation of trigger values.

Non-linear concrete behaviour has been subject to extensive research over recent decades and may be based on various constitutive models depending on the concrete behaviour that needs to be better captured. Examples are:

▦ Fracture mechanics (smeared crack model and discrete crack propagation model)
▦ Concrete plasticity
▦ Tension and rotation resistance cut-offs.

Except for the tension and rotation cut off, these models do not work with 2D beam or shell elements and require modelling of the liner with volume elements. Due to the additional effort in modelling and verification, such advanced material models are typically reserved for nonstandard design problems.

Non-linear concrete models may be simulated using one of the various constitutive laws available in the literature and commercial analysis programs. A good agreement of the simulated and the actual non-linear behaviour needs to be verified. One option is to use a base/verification model of the actual fracture-related tests (e.g. standard beam flexural tests) as per the specification and compare or calibrate the constitutive model against the test results.

11.5.2   Non-linear concrete modelling with Mohr-Coulomb failure criterion

Since tunnel analysis is usually handled through specialised geotechnical software, which may not always contain complete libraries with non-linear concrete constitutive models, a Mohr Coulomb formulation for concrete is an alternative.

The shear parameters for the Mohr-Coulomb model can be obtained by solving the Mohr-Coulomb criterion $\sigma_1 = \dfrac{2c' \cos \phi'}{1 - \sin \phi'} + \dfrac{1 + \sin \phi'}{1 - \sin \phi'} \sigma_3$ for c', using the target tensile strength of the material in the appropriate design situation for σ_3 and assuming an angle of friction σ' of between 35° and 70°.

Higher angles of friction result in a more 'brittle' reaction of the model, the actual value should be selected so that good convergence of the model is achieved. The target σ_3 can be established by conversion of the flexural tensile strength established in beam tests for CMOD1 (SLS) or CMOD3 (ULS), into corresponding direct tensile strengths. Appropriate correlations can be found – for example, in BS EN 1992-1: Eurocode 2, or the fib Model Code for Concrete Structures 2010.

11.5.3   Simplified consideration of non-linearity

Material resistance cut-off is a simplified technique to simulate the formation of cracks (simulated as loss of tensile stress transfer) and plastic hinges (loss of bending capacity and rotational stiffness). This is typically modelled in two stages:

(a) The location of the maximum tensile stress or bending moment peak is assessed – for example, through an elastic analysis.
(b) Elimination of the respective resistance at this specific location is manually assigned by the modeller.

11.5.4   Lining modelling

In general, the lining support is modelled as a beam (2D analysis) or shell (3D analysis) element where the nodes are fixed on the excavation boundary nodes of the soil. This beam or shell element can be assigned with the standard laws of elasticity, assuming an appropriate Young's Modulus reflecting the cracked or uncracked concrete sections. The section forces or stresses may then be directly calculated through the analysis and transferred into the design verification calculations. A simplified alternative is to load a shell in a structural finite element (FE) model with a distributed load provided by a geotechnical calculation.

11.5.5   Consideration of water proofing

The waterproofing types applied between the primary and the secondary support are usually either sheet or sprayed membranes. If a sheet membrane is used, the membrane-concrete interface is assumed to allow for a free shear slip between the two layers, with no transfer of tensile forces.

For sprayed membranes, depending on the type of material used, the design may model a fully bonded interface, or partial or zero shear and/or tension transfer. Consequently, depending on the waterproofing system and the associated design assumptions, the interface between the linings can be modelled as fully fixed or with special interface element formulations allowing for partial shear transfer or no shear.

For lining systems working mainly in compression with very little bending, the characteristics of the interface load transfer are not anticipated to have a substantial effect on the design and a fully fixed interface of the two layers may be considered. A fully bonded interface mobilises a composite effect of primary and secondary lining by way of shear transfer. This type of interface should not be used for long-term design situations unless substantiated by appropriate testing. The long-term behaviour of the composite effect is still subject of ongoing research, in particular with regard to creep of the material.

Modelling a fully fixed interface representing a fully bonded interface – for example, for transient design stages – is straightforward. When modelling the interface for either sheet membranes or sprayed membranes for long-term design situations, it is recommended to use specific interface elements in the design software, which may range from tension or shear cut-off formulations. They can provide a specific interface stiffness (shear and axial elastic spring stiffness between the nodes of the two layers), or they can be assigned with more sophisticated non-linear and non-elastic constitutive formulations.

When implementing complex coupling effects, the additional complexity in model validation as well as in real world testing should be considered, and an appropriate testing programme agreed with the client.

Especially for long-term load situations, creep of concrete may need to be considered. This can be realised by amending the stiffness properties of concrete in the model. Potentially different creep coefficients are applicable to the primary and the secondary lining, depending on the age of the concrete at construction and at the load applications to each layer, which can lead to the redistribution of stress.

11.6. Considering the construction process
11.6.1 Overview
Tunnel construction is a 4D engineering problem. Time, cost, hardware and human resources are the main factors that can restrict the designer to perform 2D modelling, such as axisymmetric or plane strain analysis. However, it is necessary to make certain assumptions in 2D analysis to take into account 3D geometrical effects, construction sequences and stress variations due to previous tunnelling activities. All these assumptions are to an extent subjective, and – depending on the complexity of tunnel geometry and subsoil conditions – can be difficult to estimate and validate.

If a single deep tunnel or circular shaft is to be designed and surface settlement is not of importance, an axisymmetric analysis can be employed. If multiple parallel shallow tunnels are to be investigated, plane strain analysis would be acceptable. However, in urban tunnelling – especially in areas where complex underground infrastructure already exists – the geometry of the new tunnels and their positions relative to the existing tunnels are such that presuming the plane strain condition is rarely acceptable and a 3D analysis should be performed. The complexity of the Bond Street Station Upgrade scheme, where the designer chose to model the entirety of the main construction sequences using 3D FE analysis, is indicated in Figure 11.1.

Figure 11.1 Example of 3D FEA modelling for complex situations (Bond Street Station Upgrade)

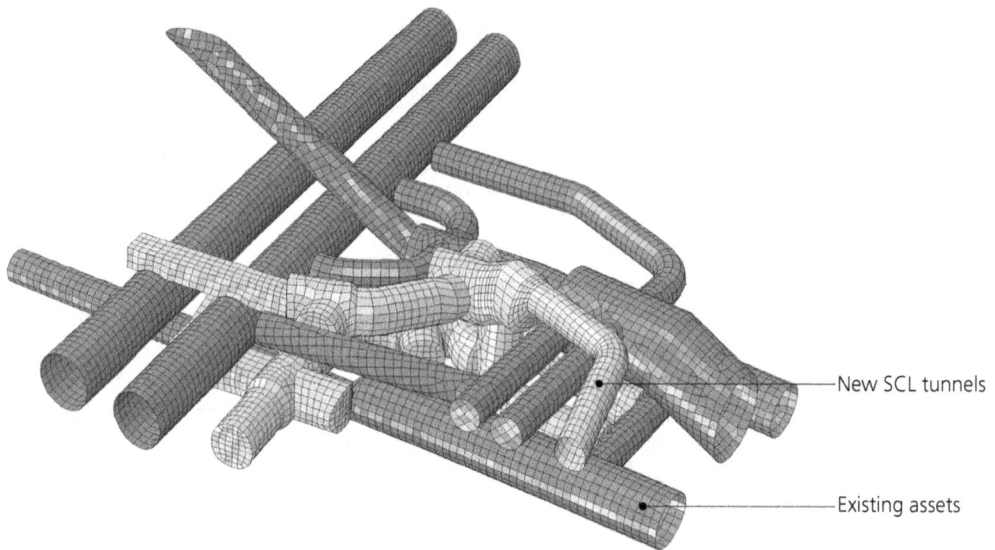

New SCL tunnels

Existing assets

Performing 3D analyses with large models is a lengthy and costly process. Trouble shooting is time consuming and sometimes the modeller might spend a lot of time on meshing and creating a large number of element sets to divide the model into individual excavation steps and the associated supports. Therefore, 2D modelling is still an attractive alternative. A 2D model should be able to mimic the ground convergence or movement occurring prior to lining installation. It can be simulated by using any of the following methods as appropriate:

- Stress method. Relaxation of internal conditions or convergence-confinement method.
- Stiffness method. Reduction of soil stiffness in the tunnel.
- Volume loss control/surface contraction method.

11.6.2 Stress method

The schematic of ground movement for a circular tunnel crown along the tunnel alignment is indicated in Figure 11.2.

A portion of the final displacement (Umax) occurs at the unsupported area (Uface) just before the tunnel lining installation. In the stress method, an internal pressure, which is a proportion of the virgin ground pressure (p_0), is applied at the periphery of the tunnel to obtain the similar ground displacement at any location between the intact ground and the completed tunnel lining with plane strain condition. Therefore, the internal pressure (p) is dependent on a so-called relaxation factor (λ) and the in situ pressure (p_0) and expressed by the equation

$$p_0 = (1 - \lambda) \cdot p_0$$

At the initial condition stage, where no excavation has taken place, a value of $\lambda = 0$ is applied; whereas when the installation of the primary lining is complete, $\lambda = 1$.

Figure 11.2 Schematic of longitudinal ground movement

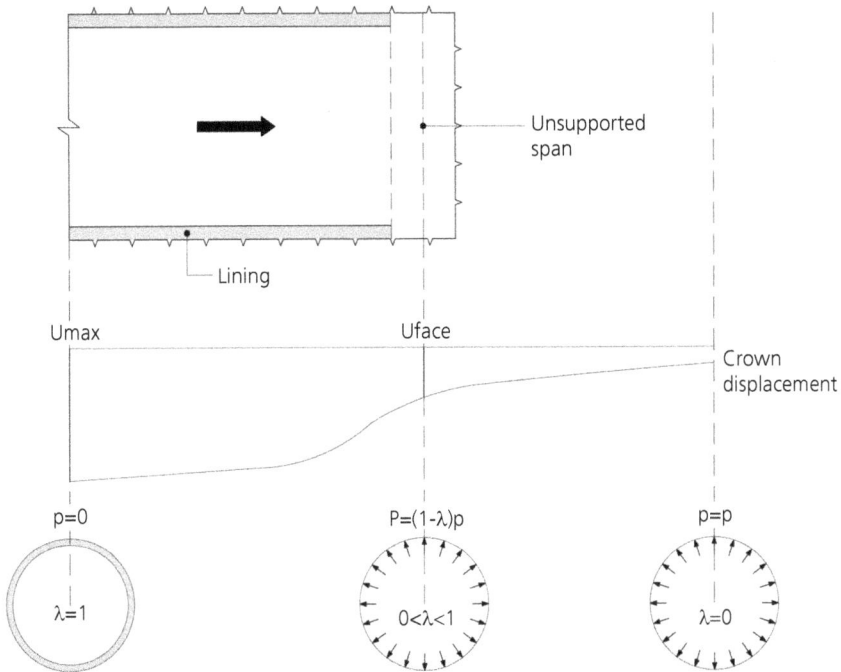

The approach is also referred to as λ-method or convergence-confinement method (Potts and Zdravkovic, 2001).

If the excavation face is subdivided into smaller sections, separate relaxation factors can be considered for individual phases.

The relaxation factor can be estimated according to Kielbassa and Duddeck (1991). Determination of the stress release ahead of the tunnel and the final ground pressure acting at the back of the lining is a soil-structure interaction problem and depends on the following factors:

- Ground stiffness
- Lining stiffness
- Unsupported span length (advance length)
- Tunnel size including excavation subdivision of the face.

The relaxation factor estimated by Kielbassa and Duddeck (1991) is based on the elasticity theory. If a relatively large plastic zone is anticipated due to high initial ground stresses and low ground strength, the relaxation factor should be increased.

As an alternative to the available closed form/empirical methods, designers may create a 3D model for the project to calibrate the relaxation factor for further 2D analyses. Also, if project data are available, where ground movements ahead of tunnel face are being monitored (e.g. inclinometers), relaxation factor can be directly used for future predictions.

11.6.3 Stiffness method

Stiffness methods employ gradual reduction of material stiffness inside the excavation area (Möller and Vermeer, 2008; Swoboda, 1979). The Young's modulus of the ground is changed by a factor inside the tunnel cross-section that leads to a stress redistribution around the tunnel. In the next computational step, similar to the stress method, the lining is installed and the stiffness of the ground in reduced to zero or the material is completely removed. Schweiger *et al.* (1997) concluded that the stress method has advantages over the stiffness method, especially in cases where excavation sequences are complex and advanced constitutive soil models are involved.

11.6.4 Volume loss control

In this method, a uniform contraction on the periphery of the tunnel cross-section is applied (see Figure 11.3), simulating a prescribed volume loss that is expected for the given tunnelling method. Following the installation of the lining, the remainder of the unbalanced force is released and equilibrium would be reached. After the lining installation additional displacement may happen depending on the stiffness of the lining system. The initial contraction should be prescribed so that when the final convergence is obtained, the total contraction equals to the prescribed volume loss. This is an appropriate method for back analysis of a tunnel problem where the volume loss has been measured.

Figure 11.3 Schematic of volume loss method

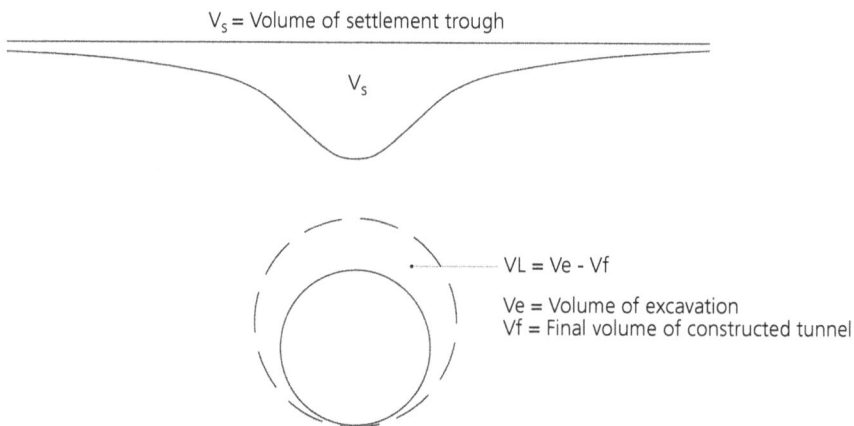

V_s = Volume of settlement trough

V_s

VL = Ve - Vf

Ve = Volume of excavation
Vf = Final volume of constructed tunnel

11.6.5 General considerations

General remarks regarding the tunnelling simulation in a 2D FE analysis are as follows:

- Initial relaxation prior to the lining installation has two major impacts: ground movement and stresses in the lining. The higher the initial relaxation the more the ground moves. This may impact the third-party damage assessment analysis. However, if a lower initial relaxation factor is applied, higher forces will be induced in the lining. In this situation, the designed lining thickness may be affected by the relaxation factor significantly. Therefore, the relaxation factor should be varied within a reasonable range to cover all of the design concerns.
- A practical method to find an appropriate relaxation factor value is to perform a simple 3D model. The point of interest would be the displacement of the tunnel crown or the

maximum settlement of the ground surface to match the 2D results with the corresponding 3D responses.

- The λ-method may be used in conjunction with all constitutive soil models. However, if an advanced constitutive model is used it is recommended not to use β-method because in advanced models where the strengths and the stiffnesses are stress-dependent the change in stiffness in the excavation area causes complications.

The FE results should be verified and calibrated against deformation measurements as soon as these data are available. Then, the relaxation factor may be updated.

11.7. Validation/calibration

The quality of numerical models, especially in geotechnics, is highly dependent on uncertainties in their input and consequently their output. The modeller should prepare a tool that fits the needs of a problem with adequate accuracy, precision and reliability. To achieve this, the modeller should fully understand the functions of the model, eliminate all errors and verify the results in the post-processing phase.

Analytical solutions can be used as a sanity-check to numerical modelling. Another sanity check is to compare results with case studies from previous projects where a tunnel is of a similar construction and in similar geology. Modelling results for new tunnels may be compared directly with the field data from existing tunnels. If the new tunnels are of different construction (e.g. a different size or shape or construction sequence) but in similar geology to existing tunnels, then the numerical modelling approach can be validated by modelling the existing tunnels and comparing the results.

Alternatively, empirical relationships derived from analysis of case histories in similar geology may be used. These often relate field measurements to tunnel depth, diameter or other factors such as unsupported length of the heading. Care should be taken to ensure that the new tunnels being designed do not fall outside of the empirical data, and that the geology and construction methods are broadly comparable.

The empirical approach is very useful in places such as London where a large number of case studies have been published. However, if a tunnel is to be constructed in geology for which there are no case studies available, empirical relationships should be used with caution, if at all.

In the pre-processing phase, a thorough check should be performed to identify any flaws in the geometric formulations and the mesh (such as wrongly assigned boundary conditions, ill meshing, or incompatible types and orders of finite elements), or in the input parameters. It is also important to increase the complexity of the model as it is being built. A substantial portion of troubleshooting in a model is associated with material (or contact) non-linearity, so a run with elastic material properties (and fixed contact interfaces) is recommended before assigning any soil plasticity.

11.8. Sensitivity analysis

Tunnelling projects are generally of a linear configuration and involve many sources of uncertainty along the tunnel alignment. Usually, there are insufficient probing and test results available at the design stage. Imperfections in construction, either in the shape of the tunnel or non-uniformity of the sprayed concrete and support elements, are common and expected. Another area of uncertainty is the natural variability or spatial variability in the case of soil parameters, which vary over space (not over time).

The designer should deal with the uncertainty involved in different aspects and take care in the selection of a constitutive model and best estimate parameters for subsoil layers. A parametric study is essential for a boundary value problem to identify the most influential parameters in the analysis. Such a sensitivity analysis could be the most appropriate method of examining and identifying such factors.

In recent years, significant advances have been made in developing various approaches to analyse uncertainty in any kind of mechanical system. Consequently, such approaches can be employed to assess the performance in a reliability analysis of a structure and ultimately the risk assessment of a project. Some of these methods have been applied in association with finite element analysis of tunnelling (e.g. see Nasekhian, 2011).

It is, typically, not possible to give only one 'exact' solution when modelling a 'real-world' tunnel due to the nonlinear and plastic effects, the time dependency of the material behaviour, the impact of local effects, and the variability of material parameters such as the sprayed concrete. Consequently, the design outcome is likely to incorporate the results of several credible design situations, selected to study the sensitivity of the system with regard to the credible range of input parameters.

11.9. Support modelling
11.9.1 Sprayed concrete lining
The primary structural component of an SCL tunnel is the concrete lining. Numerical analyses aid the assessment of soil-structure interaction and the forces in the lining. Once the loads on the supports have been calculated, the simulation of the lining can be performed using either continuum elements (also known as solid or volume elements) or shell/beam elements.

An advantage of using solid elements is that they allow for complex shapes and excavation sequences to be simulated, and they also allow more sophisticated constitutive models to be used. However, the structural section forces (Moment (M), Shear (V), Axial Load (N)) cannot be calculated directly (some algorithms are available in certain software packages) and the small thickness of the lining compared with the size of the model requires detailed meshing and usually a large number of elements.

With shell elements, the structural forces can be acquired directly, without the mesh quality being compromised. Depending on the construction method and the design approach, the secondary and the primary lining may be modelled following different individual principles. Considerations on modelling the interface of the two linings should be made, this being typically related to the waterproofing system.

11.9.2 Spiles and bolts
Anchors in rock and grouted spiles can be modelled as linear structural elements, or by improving the ground parameters in the area where these are implemented. In the first case, depending on the type of anchor used, either the start and end node of the anchor need to be related, or the entire length of the element would be connected to the soil by either fixed nodes or with the nodes using a contact-slip law. In the latter case, the area of soil where the spiles/anchors are active in

the perimeter of the tunnel should be modelled with appropriately increased material parameters (typically increased E-modulus, and/or failure limits).

11.9.3 Face stability

Face stability of tunnels is a 3D problem that can be assessed using closed-form equations, numerical methods or a mixture of both. As face stability calculations are typically a subset of the main SCL design calculation, numerical methods are mainly employed where a 3D numerical model is already available, or where large faces or a complex face geometry requires it. For geotechnical conditions dominated by large scale fracture effects (orientated joints), a wedge fall analysis should be carried out.

Closed-form methods for face stability assessments are typically suitable either for drained or undrained conditions. The designer should decide whether drained or undrained conditions need to be considered. In special cases, an undrained situation might transform into a drained one over time.

Closed-form methods for assessing face stability in undrained conditions are largely based on experimentally derived relations – see, for example, Davis *et al.* (1980). A considerable number of closed-form methods for face stability analysis in drained conditions have been published. Anagnostou (2012) gives a good overview of closed-form methods and proposes a useful equation for drained conditions based on silo theory. Kirsch (2009) has studied numerical modelling aspects of face stability in tunnelling.

As noted in Chapters 6 and 7, there are a number of options to improve face stability.

11.10. Compensation grouting

Compensation grouting involves the injection of grout into the ground between the tunnel and an overlying structure during tunnelling to compensate for ground movements. Injection of low viscosity grout at sufficiently high pressure will cause hydraulic fracture of the ground. When applied to a heavily over-consolidated clay (such as London Clay) where the minor principal stress is the vertical stress, fracture would theoretically occur horizontally resulting in thin sheets or lenses of grout being developed through the ground (Mair, 1994).

Compensation grouting can be numerically modelled in the following ways:

- Interface models. These create an interface at the level of grouting and apply fracture pressure incrementally, which can cause ground movement both upwards and downwards similar to what happens in reality (Potts and Zdravkovic, 2001). This is a rigorous method of analysing compensation grouting, but it requires considerable numerical effort if the available FE software does not offer such a feature.
- Strain-based numerical models. In this method, the fracture grouting is modelled by changing the volume of elements using an increase of volumetric strains in a series of elements at the level of grouting. The option for instant expansion of strain in a particular zone or elements is commercially available in some geotechnical software. Kummerer (2003) has reviewed these methods and compared case histories against this approach.

Both uniform pressure and interface models have been examined and compared with case histories by Kovacevic *et al.* (1996). The predicted results show that the effects on tunnel lining given by both methods are similar.

emerald PUBLISHING   ice

British Tunnelling Society
ISBN 978-1-83608-693-2
https://doi.org/10.1108/978-1-83608-690-120251014
Emerald Publishing Limited: All rights reserved

Chapter 12
Section design

12.1. Section design strategy

The tunnel lining section must be designed and executed in such a way that it will, during its intended life (including construction), with appropriate degrees of reliability and in an economical way, sustain all actions and influences likely to occur during execution and use, and meet the specified serviceability requirements.

A significant difference between tunnel design and most other structural designs is that the ground surrounding the tunnel can act both as an adverse load and as a beneficial support to the tunnel lining, at the same time. Hence, tunnel designs should be carried out in terms of a system response to a given set of external actions. Each such load/response combination should be considered separately as a design situation in accordance with BS EN 1990: Eurocode 0.

For design situations where complex non-linear modelling methods are used, load factors should be applied to the action effects rather than to the loads themselves, to allow appropriate modelling of the soil structure interaction. This approach of applying only a single load factor to the load effects prevents the use of different load factors for permanent and variable loads. The designer should represent the variability in the loads and the probability of loads occurring in combination through the appropriate selection of discrete design situations and factoring of the considered loads.

A considerable number of design situations may need to be assessed for each cross-section of a tunnel, representing different, potentially worst-case combinations of ground and water parameters. This enveloping approach should be sufficiently comprehensive to cover all relevant limit state scenarios.

In general, the structural design of SCL tunnels follows the limit state design principles set out in BS EN 1990: Eurocode 0, BS EN 1991-1: Eurocode 1 and BS EN 1992-1: Eurocode 2.

12.2. Consideration of construction sequence in the design

The construction of SCL tunnels is a cyclical process. The schematic cycle in Figure 12.1 illustrates the steps required for the installation of a typical initial/primary lining.

The task of the designer is twofold:

- Design a safe state – that is, develop a lining configuration that is in equilibrium with the surrounding ground during the planned design life.
- Develop a way to complete the circle from safe state to safe state – that is, how to build the tunnel.

Figure 12.1 The SCL construction cycle

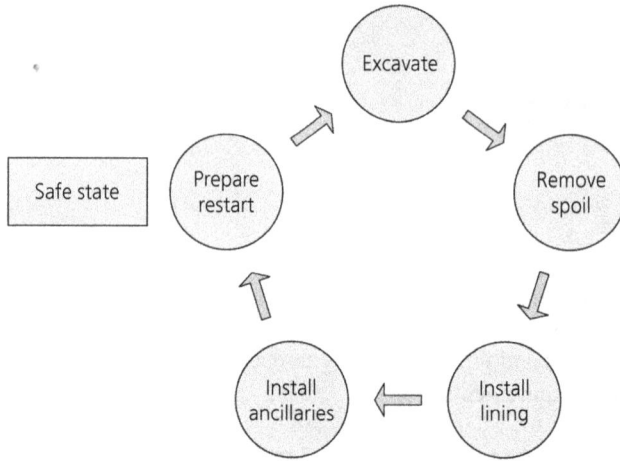

The construction process, represented by the cycle in Figure 12.1, requires a given amount of time. The ground needs to be capable of standing up, unsupported or partially supported, within this cycle time. How long the ground can stand up in a given excavation sequence and geometry depends on its geotechnical parameters and any installed advance support such as spiles, canopy tubes or lagging boards. The 'safe state' configuration needs to be developed so it can be excavated and supported quickly enough for the freshly broken ground not to collapse.

Due to the inherent variability of ground properties, a design that can be adapted to local conditions should be developed. This adaptation should be part of the design and subjected to the applicable checks and approval processes. The result can be a support classification scheme based on distinctive sets of parameters – for example, soil type, jointing, faulting, water inflow rate or similar. It is accept-able to develop separate design solutions for different design scenarios (e.g. local differences in geology), provided that these can be verified on site in order to select the appropriate support.

12.3. Design situations

The Eurocodes classify design situations into persistent, transient, accidental and seismic. For ULS design checks relevant to tunnel design, the persistent and transient design situations have identical load factors and material factors, and thus deliver an identical level of safety.

Accidental design situations can be interpreted as 'events of extreme magnitude but having a low probability of occurring' and operate on a reduced level of safety.

Seismic design situations may need to be considered for SCL tunnels, but are outside the scope of this document. For reference, refer to the publication: Seismic design and analysis of underground structures (Hashash *et al.*, 2001).

The design of a SCL tunnel contains a significant number of temporary design conditions, which can be classified either as 'transient' design situations (using the full factors of safety on the design), or 'accidental' design situations (with lower factors of safety). A design situation should only be considered as accidental if no planned man entry for this design situation is required, or if it could lead to collapse above a populated area, and underperformance or failure of the tunnel lining does not have any other irrecoverable consequences.

12.4. Material factors

Partial factors for materials, relevant to SCL design are given in BS EN 1992-1: Eurocode 2 (Section 2.4).

It might be acceptable to consider a reduced factor of safety for material parameters. This reduction can be justified due to the level of control of the relevant properties, and the implementation of control procedures to manage the risks resulting from the decreased material factor. Reference should be made to BS EN 1992-1: Eurocode 2, Annex A (informative) if this reduction is applied. Material factors for design situations are given in Table 12.1.

Table 12.1 Material factors for design situations

Design situation	Material factor for concrete g_c	Material factor for fibre reinforced concrete $g_{c,f}$	Material factor for bar reinforcement g_y
Permanent	1.5 (BS EN 1992-1-1)	1.5 (fib Model Code 2010)	1.15 (BS EN 1992-1-1)
Transient	1.5 (BS EN 1992-1-1)	*Use values for concrete design*	1.15 (BS EN 1992-1-1)
Accidental	1.2 (BS EN 1992-1-1)	*Use values for concrete design*	1.0 (BS EN 1992-1-1)

A typical approach is to define the minimum strength of the SCL prior to man entry, which is then verified by samples produced inside a designated safe zone. If such an approach is planned, an appropriate testing regime should be specified and agreed with the project team.

12.5. Action/loads

An action is a load that is applied to the tunnel structure. It can also be an effect on the structure by way of an external source – for example, change in temperature, differential settlement and moisture variation.

Actions on structures can be classified into three types, as follows:

- Permanent actions (G): ground, water, building and surcharge loads, self-weight of structural elements, finishes, fixed equipment and indirect actions such as thermal, shrinkage and any other actions that are fixed throughout the design life of the structure.
- Variable actions (Q): imposed loads.
- Accidental actions (A): events of extreme magnitude but having a low probability of occurring – for example, explosions, vehicle impact, earthquake and so forth.

Permanent and variable loads are considered in persistent, transient and accidental design situations, whereas specific accidental loads are only considered in accidental design situations. The classification of loads is not static but depends on the appraisal of the designer.

12.6.   External loads
12.6.1 Ground loads
The ground loads on the lining, established by analytical solutions or numerical modelling, would typically be considered as permanent loads. Changes in the ground loads over time – for example, by consolidation effects – are addressed by considering the changed ground loads as permanent loads in alternative persistent or transient design situations.

12.6.2   Water loads
Loads arising from groundwater are linked to ground loads due to the impact of water pressure on the effective stresses. The considerations of Section 12.6.1 apply to water loads as well.

Variabilities and uncertainties in the design water level are best addressed by considering maximum and minimum water levels as separate design situations unless identification of the governing water level (which is not necessarily the maximum water level) is straightforward.

12.6.3   Building and surface loads
Loads other than ground and water principally relate to those from adjacent structures and buildings caused by loading and in some cases, unloading. They are typically classified as permanent loads.

Building and surface loads should include consideration of future developments, and clients would normally have safeguarding limits on foundations in proximity to the tunnels in terms of allowable spatial and loading criteria. The designer should assess if the lining design complies with the client's loading criteria.

12.6.3.1 Structures with shallow foundations
Structures with shallow foundations are commonly approximated by considering a surface surcharge of 75 kPa, which also allows for future developments. The designer should satisfy himself that this is a reasonable value.

Where the surcharge load represents future development, it is usually applied with the permanent lining in place and considering long-term ground loading conditions. If a surcharge allowance is considered in intermediate design situations prior to the long-term case, it should be checked that this surcharge does not result in a less adverse stress set.

For structures with a raft or strip footings, a Boussinesq type elastic analysis can be applied to determine the anticipated additional vertical and horizontal stresses for the tunnel.

For non-trivial load patterns expected before or after construction of the SCL tunnel, it is recommended to develop a numerical model including both the foundations and tunnels to validate the design loads.

12.6.3.2 Buildings with piled foundations
Depending on the local situation, the following scenarios may need to be considered when designing for SCL tunnels close to pile foundations:

- Piles above or adjacent to, but not within the tunnel profile, where the tunnel would be subject to loads from piles. The impact on the lining should either be explicitly assessed using numerical modelling or by adopting an equivalent raft approach.
- Piles within the tunnel profile but structurally separated from the tunnel lining by a compressible element. This configuration is usually adopted where the piles are cut during tunnelling. The pile movements following cutting should be explicitly assessed and a compressible separation layer included to accommodate this movement. This approach assumes no additional loading by the cut piles onto the SCL.
- Piles within the tunnel profile and structurally connected to the lining. In this scenario, cutting of the piles is not acceptable. The tunnel lining should be designed to resist the pile loads, which are an imposed load acting directly on the lining. The impacts of cutting the piles and transferring the load onto the structure typically requires detailed assessment of both SCL and the structure supported by the piles, and a detailed design of the pile/lining load transfer structure.

Safeguarding processes and long-term asset responsibilities should be agreed with the client as part of the design process.

12.6.3.3 Unloading

Where a surface structure may be demolished or it is reasonable to expect that a future deep excavation may be undertaken in proximity to the proposed tunnel structure, the unloading induced by this activity should be analysed.

12.6.4 Surcharges

Other surcharges, such as crane loads, or loads from temporary storage, can be considered in a similar manner as buildings with shallow foundations.

12.6.5 Loads during execution

12.6.5.1 Adjacent excavations and construction

SCL tunnelling is often undertaken in proximity to other underground excavations – for example, piled boxes or other tunnels/adits/shafts constructed by tunnel boring machine (TBM) or hand mining. The impact of excavation on the tunnel lining should be assessed using numerical models and assumptions or restrictions on relative construction sequencing explicitly expressed on the drawings.

12.6.5.2 Construction surcharge loading

Surface surcharge loads are not generally applied to the construction stages. However, for shafts it may be necessary to consider the surface surcharge due to construction plant. This is often taken as 20 kPa to account for typical construction plant loading. The designer should satisfy themselves that this is suitable and take note that local horizontal loads onto the structure may be higher closer to the surface, where uniform pressure and hoop loads are lower.

12.6.5.3 TBM loads

At the interface between TBM and SCL tunnels – for example, reception chambers – an SCL headwall is often formed. The headwall design should take account of the predicted pressure from the TBM as it approaches the headwall.

TBM launch and assembly/disassembly loads are often applied to SCL launch and reception chambers and these loads can be significant.

12.6.5.4 Compensation grouting

Compensation grouting is a well-established technique employed on tunnelling projects to minimise settlement. It works by injecting grout into the ground to compensate for the ground loss at the tunnel face and the following installation of a relatively flexible lining. As compensation grouting overcomes the overburden pressure, the reaction to these pressures provided by the ground and tunnel lining below the grouting horizon can be significant. Important factors include: grouting array depth, building mass and cover from the tunnel crown to the array level. If there is sufficient cover between the array depth and the tunnel crown, compensation grouting may not need to be considered in the SCL design, both in terms of loading to the lining and no grouting zone criteria.

The lining forces are usually assessed post ring closure – that is, on the completed lining – and an exclusion zone is applied behind the face to ensure grouting pressures are not applied to the partially constructed lining. A typical no grouting zone is shown in Figure 12.2.

Figure 12.2 Typical compensation grouting exclusion zone

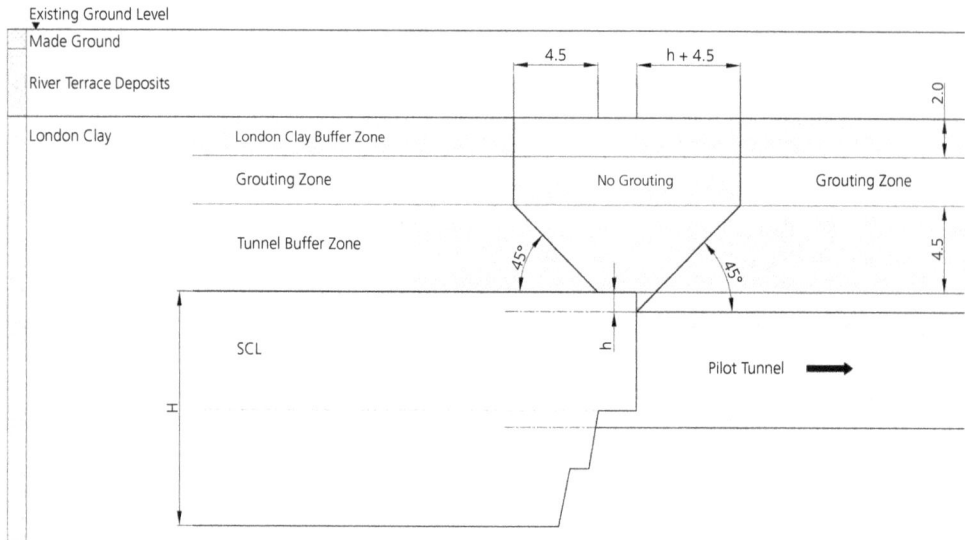

The effect of compensation grouting on tunnel linings can be analysed by numerical modelling, as described in Chapter 11 (Section 11.10).

12.7. Internal loads

12.7.1 Fixing loads

The secondary lining may be subject to loads from internal equipment – for example, fans, dampers, suspended floors, escalator truss lifting and so forth. A critical situation is often following secondary lining installation but prior to any external ground loading being applied to the lining; when the lining is not subject to any beneficial axial loading.

Depending on the intended use, fixing loads can be permanent or variable. The designer should classify the respective loads and, if necessary, normalise the design loads.

12.7.2 Thermal loads

Thermal loads can be classified into: effects resulting from the hydration heat generated by the concrete, effects from ambient temperature, and fire loads:

▨ Heat generated by hydration leads to thermally induced strains of the concrete acting against restraint provided by, for example, the ground or other concrete layers. The long-term shrinkage of the concrete is typically released by cracking. Thermal loads from hydration heat are commonly only considered in SLS design. Further guidance on this matter is given in CIRIA (2018) *C766 Control of Cracking Caused by Restrained Deformation in Concrete*.
▨ Ambient temperature effects result from the difference in temperature between the tunnel intrados and the surrounding ground. While the tunnel intrados temperature is specific to its use over the course of the year, the temperature of the surrounding ground can be estimated as between 8°C degrees and 11°C and remains stable unless influenced by external factors.

Designing for fire is discussed in Chapter 13.

12.7.3 Traction loads

If considered significant, the self-weight of the train and braking/acceleration loads should be included in the design.

12.7.4 Internal pressure

Internal pressure can result from high-speed trains, or filling pressure of hydraulic tunnels, or blast effects, or compressed air.

12.8. Accidental loads

12.8.1 Impact loads

Crash or derailment loading onto the tunnel lining is not typically considered in the analysis as a stability issue and is therefore a local damage issue for the tunnel lining that should not require analysis. The exception is headwalls or tunnel bifurcations. For internal structures such as platforms, crash loading should be considered.

12.8.2 Blast loads

Where specified, blast loads should be considered by the designer. Tunnels can generally mobilise adequate passive resistance of the ground to contain blast effects. Blast loads should always be considered at the most disadvantageous location unless specified otherwise. Such locations are often at peaks of bending moments or shear forces where loss of section capacity can lead to lining failure.

Dynamic blast loads are often approximated with equivalent static loads. Where this is not the case, specific modelling techniques are required, and these are outside the scope of this document.

12.9. Load combinations

12.9.1 Design for the ULS

For the definition of ultimate limit states refer to Eurocode 0, 6.4.1. The STR ultimate limit state is typically used for the design of tunnels. The load and combination factors relevant for this design are defined in the National Annex (NA) to BS EN 1990: Eurocode 0. Commonly, Set B as per Table NA.A1.2 (B) is used together with Table NA.A1.1 for the design of underground structures.

For ULS tunnel design for permanent and transient design situations, the following load factors taken from BS EN 1990: Eurocode 0 are recommended.

Where specific accidental design situations are to be assessed the designer should refer to the specific clauses of the corresponding design code:

- For unfavourable permanent actions: $\gamma_{G,sup} = 1.35$
- For favourable permanent actions: $\gamma_{G,inf} = 1.0$
- For unfavourable variable actions: $\gamma_{Q,sup} = 1.5$
- For favourable variable actions: $\gamma_{Q,inf} = 0$

Combination factors for variable actions in ULS are typically as follows:

- For non-fire temperature actions: $\psi_{0,temp} = 0.6$
- All other transient actions: $\psi_0 = 0.7$

In general, the actions that have been identified as relevant for the design should be grouped into appropriate load combinations. These load combinations represent typical design situations occurring during the design life of the tunnel. Design situations should be varied enough to capture credible variations in input parameters. For example, a design situation might represent the loads resulting from assuming an upper bound ground water level, whereas another design situation might represent the same design cross-section at a lower bound ground water level. Similar considerations can apply to temperature profiles that often consider a 'summer' and a 'winter' temperature profile.

Where design codes such as the Eurocodes require different partial factors and combination factors for different actions (such as permanent and variable actions), it is recommended to factor the actions so that a single load factor can be applied to the action effects. Typically, the factoring should be against the load factor used for permanent loads as these are likely governing. For instance, if partial factors for a variable load and permanent loads are 1.5 and 1.35, respectively, a surface surcharge load (as a variable load) is multiplied by '1.5 divided by 1.35'. Once the effect of actions is calculated, they are factored by the permanent load partial factor (i.e. 1.35) only.

The load combination for permanent and transient ULS design situations is described in BS EN 1990: Eurocode 0 (Section 6.4.3), which describes the application of load factors and combination factors.

12.9.2 Combinations for SLS design

The design for serviceability limit states is likewise undertaken for load combinations, notably the quasi-permanent, the frequent and the characteristic combination. Typical design checks linked to SCL tunnels such as deformation checks and crack width calculations use the quasi-permanent load combination as per BS EN 1990: Eurocode 0 (Section 6.5.3 (6.16b)). Values given below assume the design is for the quasi-permanent load combination.

Similar to ULS design, the designer should group actions occurring concurrently into representative design situations. Permanent actions relevant to the design situation are considered at unity, whereas all relevant variable actions are considered with a combination factor ψ_2.

Table NA.A1.1 provides different values for ψ representing different categories of buildings. Regarding the combination factor for the quasi-permanent load combination ψ_2, a higher value is typically associated with the expected frequency of occurrence and the proportional relevance of the transient action. The ultimate selection of the appropriate combination value should be made by the designer on the basis of the parameter range defined in the codes.

For SLS tunnel design in the quasi-permanent design situation, the following load factors are recommended:

- For unfavourable permanent actions: $\gamma_{G,sup} = 1.0$
- For favourable permanent actions: $\gamma_{G,inf} = 1.0$
- For unfavourable variable actions: $\gamma_{Q,sup} = 1.0$
- For favourable variable actions: $\gamma_{Q,inf} = 1.0$

Combination factors for variable actions in SLS are typically:

- For non-fire temperature actions: $\psi_{2,temp} = 0$
- All other transient actions: $\psi_2 = 0.6$

12.10. Plastic redistribution
At highly stressed areas of an SCL cross-section, the concrete should generally be considered as a cracked material, where the resistance of the cross-section against bending moments is reduced. This leads to a redistribution of bending moments away from moment peaks to less utilised parts of the cross-section.

This consideration can be made either explicitly by using advanced material models, through definition of discrete concrete hinges, or implicitly by reduction of the section modulus.

Guidance on plastic analysis is contained in BS EN 1992-1: Eurocode 2 (Section 5.6). For SCL design, the kinematic method (plastic hinges) is the most relevant approach. The code defines minimal requirements that must be met to omit the explicit rotational capacity check. However, these conditions are typically not complied with in SCL design and, therefore, the rotation of the plastic hinge requires to be checked against the code requirements.

The basic value of allowable rotation $\theta pl,d$ is presented for Class B and C reinforcement with their associated characteristic strain at maximum force $\varepsilon uk \geq 5.0\%/7.5\%$. It is not recommended that Class A reinforcement (characteristic strain at maximum force $\varepsilon uk \geq 2.5\%$) is used for design where plastic hinges can develop, as similar criteria have not been developed in standards or quasi standards.

For fibre reinforced concrete (FRC), it is proposed that the maximum allowable rotation is derived using analogies to constitutive tests such as the BS EN 14651:2005, where the plastic rotation capacity is described. The test is carried out to a CMOD ≥ 4.0 mm on a cross-section height of $h=125$ mm (25 mm notch in a 150 mm beam opposite the central loading point as the transducer is placed on the surface of the beam). Using the relation of Figure 5.5 in BS EN 1992-1: Eurocode 2, as presented in Figure 12.3, the tested rotation is taken as

$$\theta_{s,EN14651} = \tan^{-1} \frac{4.0 \ mm}{0.6 \cdot h = 150 \ mm} = 44.42 \ mrad$$

Figure 12.3 Plastic rotation of reinforced concrete sections for continuous beams and continuous one-way spanning slabs

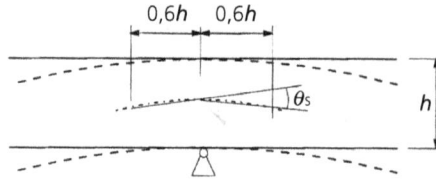

Assuming an associated ratio of x_u/d of approximately 0.4 following BS EN 14651:2005 Annex A, the associated allowable rotations for Class B and Class C reinforcement would be in the region of 5–10 mrad (BS EN 1992-1: Eurocode 2, Figure 5.6N). It is therefore considered that using the Class B curves from Figure 5.6N represents a reasonably conservative assessment for the assessment of the rotational capacity of FRC, provided that the BS EN 14651:2005 beam test has been conducted successfully to the required CMOD of 4 mm.

British Tunnelling Society
ISBN 978-1-83608-693-2
https://doi.org/10.1108/978-1-83608-690-120251015

Chapter 13
Designing for fire

13.1.   Fires in tunnels

Examples of major fires in tunnels in recent history (Channel Tunnel, Mont Blanc etc.) have demonstrated that tunnel fires are low-frequency, high-consequence events, that can cause multiple fatalities and injuries. Following a fire incident, the cost impacts of both repair and downtime can be extreme. In the case of road and rail tunnels, the socio-economic impact of a major artery or community access being closed for repair works over an extended period can have local, regional and national impacts. Smoke from fires can travel along a tunnel, resulting in a toxic atmosphere that can be fatal, and reduced visibility, making it more difficult to escape. A coordinated approach between designers, emergency services and operators is required to develop the structural design, ventilation strategy, escape procedures and determine whether places of safety are needed.

Tunnels do not typically collapse following a fire incident, but if the heat is severe and sustained for a significant length of time, the thickness and structural capacity of the lining would steadily reduce, and collapse of the tunnel would become a possibility. This would be of concern for tunnels in ground that is not self-supporting, and/or under water or near critical surface assets. These factors should be considered in a risk-based approach for fire protection.

13.2.   SCL response to fire

Concrete, including sprayed concrete, is non-flammable, and not a potential fuel source for a fire. However, concrete exposed to a fire will tend to spall, where free water trapped in the concrete matrix heats and evaporates, building up pressure that forces the concrete surface away, progressively reducing the concrete section. Above 300°C, concrete can also experience chemical changes, as the loss of water causes shrinkage, modifying density and the bonding forces between the crystals of the cement gel, and accordingly a loss of strength. This strength loss is irreversible, and any significant fire will likely lead to a requirement to repair the tunnel after the event.

Sprayed concretes typically use fine aggregate; sands and up to 10 mm size gravels, sometimes with additional fines in the form of limestone powder. The water–cement ratio is kept as low as possible, typically 0.4–0.5, and the result is a dense impermeable mix. These conditions are prone particularly prone to the spalling risks described above, sometimes explosively.

For general information on design of tunnels for fire, refer to the BTS *Tunnel Lining Design Guide* (2004: Section 4.6) and PAS 8810:2016 (BSI, 2016).

BS EN 1992-1-2: Eurocode 2, Part 1–2: General rules – Structural fire design deals with the design of concrete structures for the accidental situation of fire exposure and considers passive methods of fire protection.

Depending on the fire load, the following considerations apply:

▦ If the design fire is based on the ISO 834 temperature-time fire curve, the structural fire design can be directly undertaken on the basis of BS EN 1992-1-2: Eurocode 2.
▦ The effects of other fire curves, such as the EUREKA curve (also known as the EBA fire curve or EUREKA EN 499 – a severe fire design curve used primarily for assessing the fire resistance of rail tunnel structures), can be assessed on the basis of BS EN 1992-1-2: Eurocode 2 by calculating temperature distribution in the member, the resulting additional loads on the lining, and the temperature dependent material resistance for a number of timesteps.

The client may request a fire resistance that leads the designer to a given minimum lining thickness, cover and material selection to ensure the necessary requirements can be met.

13.3. Protection options and considerations

Protection options that should be considered include the following: including a fire suppression system within the tunnel (sprinklers, deluge system etc.), protecting the structure using cladding or coatings, or allowing the concrete to release the water pressure using micro-synthetic fibres.

13.3.1 Fire suppression systems

If activated quickly, fire suppression systems can stop a fire from building in intensity, lower the temperatures and prevent spread. Known as 'active protection measures', these systems assist with getting users out of an incident tunnel. Concerning tunnel design, the main factors are as follows:

▦ Spaceproofing for the system.
▦ Providing adequate drainage and sumps.
▦ Treatment facilities to prevent fire polluted water from entering the local environment.

Drainage systems should be designed to allow for the extinguishing of liquid fuels entering the system. The locations of fire compartments should be coordinated with ventilation and escape route design.

13.3.2 Cladding and coatings systems

Fireproofing materials come in the form of cladding systems or spray-applied coatings or paints. These can be selected to suit the fire load scenario for the project and can provide sufficient protection to the underlying structure that it is minimally impacted by the heat and does not exhibit spalling or strength reduction. Spaceproofing, loading, surface roughness and other aesthetic criteria should be considered when selecting these options.

If a material is to be bonded to the surface of the tunnel there may be additional activities required to prepare the surface and it may be necessary to combine this with mechanical bonding through the inclusion of a mesh and fixings. If the final coating on an SCL tunnel is the fireproofing, this is the surface that receives items subsequently fixed to the tunnel. If the coating is not structural or insufficiently bonded to the substrate, the length of fixings should be prescribed for future users to tie back into the structural concrete and tolerances should be accounted for.

While potentially bulkier than a spray applied system, there may be benefits to using a thermal barrier cladding system, and it may be possible to coordinate this with the architectural vision, or to facilitate using a sprayed secondary lining without the need for smoothing or casting a final surface. Fire protection measures can restrict access to the structural lining, making it harder to inspect, but may have benefits for maintenance in terms of cleaning and providing a better surface finish that reduces lighting requirements.

Fireproofing cladding panels, offset from the tunnel surface, should be designed for wind and buffeting loads, and impacts from collisions. The cladding system may need to be replaced within the tunnel lifetime, adding to the whole life costs, not only in materials but also any shut-down period required to complete such work. Cladding panels are factory-controlled products with extensive testing, and might open the possibility to increase the level of fire protection later without having to carry out major structural upgrade work.

13.3.3   Micro-synthetic fibres

Micro-synthetic fibres (monofilament polypropylene fibres) can be added to sprayed concrete to control spalling. The fibres melt at comparatively low temperatures, approximately 160°C, leaving behind tiny channels that allow any trapped water vapour to escape without fracturing the concrete. As with the thermal barriers, the fibres are a passive form of protection, although in this case in the event of a fire the surface of the concrete may need to be replaced.

Test data for dosages of micro-synthetic polymer fibres that can be included in an SCL mix can be found from manufacturers, although this may need to be supported by testing of the chosen mix and specified fire load. Previous testing suggests that between 1–2 kg of polypropylene fibres per cubic metre of concrete can provide adequate fire resistance and anti-spalling properties in the event of a fire. The micro-synthetic fibre dosage is related to the type of fibre as well as the concrete mix design and fire load.

In addition, testing suggests that a 50 mm thick regulating layer containing monofilament polypropylene fibres and calcareous type aggregates can provide passive fire protection to an SCL composite lining subject to a hydrocarbon fire (King, 2018).

emerald PUBLISHING   ice

British Tunnelling Society
ISBN 978-1-83608-693-2
https://doi.org/10.1108/978-1-83608-690-120251016

Chapter 14
Instrumentation and monitoring

Monitoring is an integral part of the SCL construction process and a key element in risk management. Guidance on instrumentation and monitoring is given in the BTS (2011) *Monitoring Underground Construction: A Best Practice Guide*. Instrumentation and monitoring is also discussed in the BTS (2004) *Tunnel Lining Design Guide*. The specification for monitoring, recording and the control process for underground works is described in the BTS *Specification for Tunnelling*. Another reference is the ITA Report No 009/Nov 2011 – *Monitoring and Control in Tunnel Construction* (ITA, 2011).

The key objectives of monitoring in SCL construction are as follows:

- To provide data for making informed decisions as an integral part of construction activities.
- To provide assurance to the client and to third parties that a project is being delivered in accordance with the project requirements and any legal agreements.
- To verify or validate design assumptions or predictions.
- To provide evidence that materials and workmanship comply with requirements.
- To provide data for risk management.
- To provide data that may be used to trigger pre-planned contingency actions to control risks associated with the works or existing assets.
- To provide data for research and design development.

Monitoring is typically provided both within the tunnel or shaft being constructed as well as outside the underground structure being built, in adjacent existing infrastructure, within the ground, or at ground surface level.

14.1. Optical 3D monitoring

The principal instrumentation equipment used for monitoring underground structures from within the SCL tunnel or shaft is the use of optical 3D monitoring and geodetic levelling equipment including robotic total station equipment. Displacements are monitored by measuring absolute movement in all directions (vertical, horizontal and longitudinal) of convergence targets fixed to the tunnel lining.

Monitoring targets should be installed and read as soon as it is safe and practicable to do so, following the spraying of the lining, to establish baseline readings. Targets should be read from a point of safety. The frequency of readings should be agreed with all relevant parties but typically readings should be taken every 24 h for the first two weeks, followed by two readings per week for the next four weeks, followed by one reading per week thereafter until agreed by all parties that readings can be reduced.

Instrument arrays should be provided at regular intervals (typically every 10 m) and additionally at specific areas of concern such as at openings or at critical sections. Redundancy and diversity should be considered when designing instrumentation to provide continuity of data should instrument failure occur.

The deformation of inverts, particularly flat inverts in large caverns, needs to be monitored carefully. This is usually done with levelling studs in the tunnel invert, and it is essential that these are protected to prevent damage by the temporary backfill and expected traffic loads.

14.2. Tape extensometers

Tape extensometers can be used to measure convergence, but are not usually recommended as they can put operatives in a hazardous location within the tunnel environment, and similar results can be obtained from more modern remote measurement without the need for working at height or in hazardous locations.

14.3. Pressure cells

Two types of pressure cells are sometimes specified for SCL construction:

- Radial (sometimes referred to as boundary) pressure cells measure the pressure between the extrados of the SCL and the ground.
- Tangential (sometimes referred to as axial or embedded) pressure cells measure the stress within the SCL in the circumferential direction.

There is some debate in the industry concerning the effectiveness of pressure cells. Some have a limited appetite for installing them; claiming they are time consuming and difficult to install and, in normal site conditions, the information they produce is generally unreliable and often ignored. However, if properly installed and interpreted, very good information can be obtained (Jones *et al.*, 2023) about the development of stresses on and in sprayed concrete linings in both the short- and long-term, that it is not possible to obtain by any other means.

14.4. Strain measurements

Convergence monitoring has traditionally been carried out using high precision total stations capable of measuring 3D coordinate positions of prisms or reflector targets fixed into the tunnel lining. Monitoring is carried out manually by surveyors working within the tunnel. Best practice accuracy is ±2–3 mm, but can often be worse than this. Targets are also frequently knocked or damaged by construction activities.

Strain gauges may be embedded within the sprayed concrete to provide continuous monitoring of strain in the sprayed concrete lining. These may be individual strain gauges, typically of the vibrating wire type, or can consist of an optical fibre.

Optical fibre strain monitoring (OFSM) is a recent innovation in SCL tunnelling and has been used at Bond Street and Farringdon Crossrail stations. Fibre optic arrays can be placed in underground excavations at locations determined by the design team, often at the convergence monitoring locations to enable comparisons to be made. The fibre optic arrays can be manufactured to suit the specific tunnel geometry such that the continuous fibre optic cable has a sensor within it at the correct clock position around the excavation heading.

The installation process does not differ significantly from pressure cells. After the application of the initial sealing layer (typically when a specified early concrete strength is reached, releasing the exclusion zone), the cable is pinned to the intrados. The primary lining is then sprayed, fully encapsulating the cable, except for the outgoing signal cable linked to the interrogator unit. The signal cable is temporarily fixed to the tunnel lining and sends continuous data back to the interrogator unit. The data can be made available in real time. The frequency of the readings can be configured to suit project needs, but once every 15 min is typical.

The arrays provide real time structural monitoring with continuous and accurate measurement of strain (to within 1–2 micro-strain) and temperature (to within 0.1 degrees), and the data can be hosted on a secure website for remote interrogation. It is essential to monitor both temperature and strain independently.

It should be remembered that strain is not in itself a useful measurement, because it includes plasticity, creep, shrinkage and thermal effects, all of which have a similar magnitude to the strain due to loading. The amount of strain at which failure of the concrete will occur depends on timing of loading increments relative to the maturity of the sprayed concrete, as well as environmental factors such as temperature and humidity. It is possible to use the 'rate of flow method' (England and Illston, 1965) to back-calculate stresses from strains, but that requires knowledge of the plasticity, creep, shrinkage and thermal behaviour of the concrete, and how these behaviours vary with age. An example of such a calculation for a sprayed concrete lining, with comparison to pressure cells, is given in Jones and Grand (2024).

The design life of strain gauges or OFSM typically exceeds the construction time of the tunnel, and may extend for some decades into the service life of the tunnel. It is recommended to plan for the integration of the OFSM into any planned structural health monitoring of the permanent asset.

14.5. Strength testing

The development of concrete strength is established through on-site and laboratory testing and compared against the strength development requirements (see Chapter 8: Figure 8.2). Methods are outlined in BS EN 14488-2:2006.

At early ages, a calibrated gauge is used to record the force required to drive a needle into the concrete. This is used when the concrete is still weak, so a safe place to carry out the test – often a panel that can be stored away from the exclusion zone at the face – is used. As strength develops, this method is replaced by threaded studs driven into the concrete, with the force to pull them out recorded. As strength increases further, cores are taken and tested at a laboratory. These cores can be taken directly from the lining, if appropriate, or again from specially created panels using the same mix at the same time as the lining is sprayed.

The development of concrete strength should be reported back to the daily review meeting, to demonstrate that the material is performing in line with requirements.

14.5.1 Thermal imaging

Thermal imaging, coupled with software using thermodynamic parameters obtained from laboratory testing and thermomechanical parameters obtained from on-site calibration, can also be used to monitor the early strength development of sprayed concrete. Temperature is the most important factor affecting the strength development of a given concrete mixture, with rate of strength gain approximately doubling every 10°C.

A maturity method, commonly used for precast or in situ concrete, has been adapted to be used for the more complex situation of accelerated sprayed concrete, such that the compressive strength of the sprayed concrete may be determined in real time at the tunnel face using specialised software (Ahuja and Jones, 2016; Jones *et al.*, 2014, 2017; Weiher *et al.*, 2017). This method is known as 'SMUTI' (strength monitoring using thermal imaging).

SMUTI uses a thermal imaging camera to measure the temperature of the sprayed concrete lining, which means that the strength can be determined from a safe position within the tunnel several metres back from the face. It also means that the whole of the sprayed concrete lining can be monitored, not just a discrete location representing a very small volume of sprayed concrete.

14.6. External monitoring

External monitoring consists of all the monitoring installed outside of the tunnel. This can include extensometers, inclinometers and piezometers installed in boreholes, as well as surface settlement monitoring and third party asset monitoring.

External monitoring has the advantage that measurements can be taken prior to underground excavation and thus can capture any ground movements or changes in pore water pressure prior to, during and after the construction of the underground structure. For external monitoring, baseline measurements should be taken to establish stability of monitoring system and existing diurnal, seasonal or tidal fluctuations. External monitoring should be analysed and interpreted in conjunction with in-tunnel monitoring.

14.7. Setting trigger values

The methodology for determining the trigger levels should be part of the risk management process and be based on the specific design and construction methodology for a particular area of underground construction. Trigger levels should be used as a warning mechanism, enabling preventative measures to be introduced in an acceptable timeframe. It is normal practice to establish trigger values for key indicator values such as displacement, tilt, strain or pressure.

There are a number of ways in which trigger levels can be configured but it is important that the trigger values are well defined and understood by all relevant parties and be appropriate to the structure being monitored. For SCL, it is typical for displacement trigger values to be used and set in terms of linear or angular movement or the rate of change of movement as appropriate. When measuring internal tunnel displacement, it is important to take account of an estimated percentage lining displacement occurring prior to installation and first reading of a monitoring point.

When setting up trigger levels, the following should be specified:

- Procedures for passing on information.
- Allocation of responsibilities between the key parties.
- Time allowed for passing on information or decision making.
- Remedial actions or responses for dealing with foreseeable situations.

Typically, there would be three trigger levels: Green, Amber and Red, as indicated in Figure 14.1.

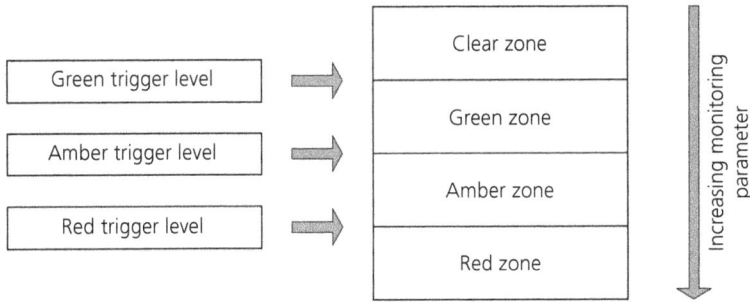

Figure 14.1 Trigger level categorisation

Typical definitions of the zones and trigger levels with associated actions are described below.

Clear Zone – sometimes a Clear Zone is used, and it is the zone when monitoring is initiated, prior to any trigger levels being reached.

Green Trigger Level – this trigger alerts that the Green Zone has been reached. In the Green Zone, the monitored values are less than the Amber Trigger Level. The construction work is considered to be in a normal and safe condition. In SCL design, the Green Trigger Level can be set at a percentage (often about 70%) of the predicted displacement.

Amber Trigger Level – this trigger alerts that the Amber Zone has been reached. The Amber Zone is reached when the monitoring/review process discovers results exceeding the Amber Trigger Level or trend rate. The Amber Trigger Level can be set at the displacement predicted by the designer. The Amber Trigger can indicate a potential problem but should not be of sufficient severity to require cessation of the works. The action should be to check instrument function, visually inspect the structure being monitored, and review the frequency, type of monitoring, the construction methodology and the modelling assumptions. This should all be done within 24 h.

Red Trigger Level – this trigger alerts that the Red Zone has been reached. The Red Zone is reached when the monitoring/review process discovers results exceeding the Red Trigger Level or trend rate. The Red Trigger Level should be set below the ultimate capacity of the structure. This level may equate to a displacement above which unacceptable damage is expected to occur. The action should be an immediate check on instrument function, and a visual inspection of the structure being monitored. It should also initiate a pre-determined response that may include a controlled cessation of the tunnelling process, a back analysis of the event and a modification to the design and construction process. An emergency review meeting should be held with relevant parties. The designer should relate the trigger level to the ultimate capacity of the structure to provide context to the trigger levels, and to assist in the decision-making process and emergency response.

Black Trigger Level – sometimes a separate Black Trigger Level is added where a full emergency response would be initiated and/or third-party asset owners informed.

It should be noted that Green, Amber, and Red Trigger Levels for external monitoring are often not comparable and may require a different response in the event of a trigger level breach. The inter-relationship between these should be understood and documented to avoid confusion.

The response to trigger values is an important part of the risk management of underground construction. It is essential that at the outset of the works the designer and the construction team fully understand the trigger levels and what they represent for each construction stage. An action plan should be provided that includes all the pre-planned contingency measures that should be taken if a trigger value is exceeded.

Considerable care should be exercised in defining what the in-tunnel trigger levels represent, and to differentiate between in-tunnel and external trigger levels. The project teams (and affected stakeholders) should be properly briefed in relation to the trigger levels. It is recommended that emergency simulations are undertaken, by the project teams, to prove their operational readiness should triggers be breached.

14.8. Understanding and interpretation

Underground structures in soft ground with relatively shallow cover but with stiff linings often have low values for trigger levels. Initial movements are often rapid and may approach predicted trigger levels but then stabilise as soon as the lining ring is completed. This demonstrates that monitoring results require interpretation before use, and that a purely deterministic approach may not to be conducive to safe and efficient tunnelling. The implications of the monitoring output values need to be understood and should be reviewed in conjunction with:

- Progress and stage of construction
- All other monitoring (piezometers, surface monitoring, inclinometer, extensometer, pressure cells)
- Sprayed concrete properties (both wet and hardened)
- Geotechnical conditions
- As-built thickness and profile control
- Visual examination
- Unusual occurrences
- Scatter of results, accuracy of readings and potential external influencing factors.

Traditionally, tunnel linings are designed based on stresses but monitored based on displacements. There is no easy way to relate trigger levels based on displacements to stress against capacity (also known as 'utilisation'). The key is often to look at the trending of displacement rather than the absolute values (Rokahr et al., 2002). By considering the relative movements of the measuring points, an approximation of the strain in the lining can be assessed and used as a guide in determining the adequacy of the lining (Macht et al., 2003; Rokahr and Zachow, 1997, 1999; Stärk et al., 2001).

Trigger values need to be selected with care such that adequate warning is given of any impending failure or damage to assets but not such that too restrictive values for trigger levels unnecessarily impact the progress of the works. Attention should also be paid to likely environmental fluctuations and the accuracy of the monitoring system employed – if triggers are likely to be exceeded by these factors then either a more accurate means of monitoring needs to be used or the triggers need to be increased.

emerald
PUBLISHING

ice

British Tunnelling Society
ISBN 978-1-83608-693-2
https://doi.org/10.1108/978-1-83608-690-120251017

Chapter 15
Sustainability in design

When considered under a three pillars model (assessing social, economic and environmental benefits), tunnels are often part of an overall solution that has sustainability benefits – such as public transport networks, power distribution or wastewater management. As part of any engineering design, questions should still be asked about whether a tunnel is needed, whether its route can be optimised, and whether the solution proposed offers the best benefit possible when viewed holistically against other options; including reducing demand or reusing/repurposing existing assets.

These big questions are important as, when viewed with respect to carbon emissions, tunnels can be seen to use large amounts of carbon rich materials and have significant room to improve performance. As noted in the ITAtech report (2023) *Low Carbon Concrete Linings*, it is the profession's responsibility to address this issue going forward.

This need to improve is particularly true of sprayed concrete linings, where design thicknesses can be significantly larger than for tunnels of equivalent dimensions supported using segmental linings. This is understandable in part, due to the in situ application processes and allowances that should be made for variability, that factory-created support can avoid. However, designers should also look to see whether they are optimising their lining designs and making the most of the materials, equipment, analysis methods and engineering judgement available.

Following the collapse of the tunnels at Heathrow, safety processes in SCL tunnel improved greatly. The daily review of excavations, as-built support, monitoring, ground conditions and materials can make it one of the most controlled construction processes in a project. At the same time, lining thicknesses in the years after may have also increased, potentially due to an idea that with a thicker lining there was less chance of collapse.

For every cubic metre of concrete required in a theoretical SCL design, data from construction has shown there can be a further cubic metre of concrete brought to site. This can be accounted for in:

- Materials discarded for being out of specification
- Overbreak
- Construction tolerances
- Face sealing and other temporary works
- Material lost in lines, starting and finishing pumping and spraying
- Use of sprayed concrete for pre-construction trials and for test panels during construction.

Some of these items are required to respond to the risks of open faced, flexible shaped tunnelling. Others can be avoided/minimised or managed, and should be.

15.1. Carbon calculations

During the design process, many projects require the designer to input to a carbon model, and look to see improvement as the design progresses. It is important that calculations include the wastage factors noted above, at least until solutions that reduce these values can be incorporated. Comparison should be made based on standardised figures that reflect the location of the project and availability/proximity of resources.

It is important to note that proposing a very conservative design at early design stages, and then coming back to a more standard or best practice, design at detailed design does not reflect a true carbon saving. Continuing with an overly conservative design would be more wasteful, but to actually take steps forward engineers need to be starting from a position of current best practice and bettering it each time they are able. This can be difficult when a project sets carbon reduction targets as percentages against initial estimates, rather than benchmarking against appropriate comparative projects.

15.2. Lining configuration

As noted throughout this guideline, the SCL designer must choose a lining configuration that meets all of the project requirements. These decisions should also be looked at through the lens of sustainability:

- Can the internal spaceproofing be optimised based on the lining solution adopted?
    - o Can internal layers of lining be removed?
      If the internal structure will be fireproof or waterproof, does there need to be a secondary lining? Would adopting an internal system be more efficient over the full life cycle of the project? How will it be supported and how often would it need to be replaced?
    - o Can internal cladding be removed?
      If the secondary lining can meet performance requirements, can you remove the space and materials allocated to cladding? Will the public be exposed to the tunnel lining? Will they mind seeing the supporting materials?
- Do you need all the layers?
    - o Sealing layers applied to the ground can be an important feature if personnel access is required into the face during construction. However, if the design has avoided this requirement, can the thickness allocated to this be removed? In some cases, sealing layers are applied to address long term durability requirements, based on aggressivity of the ground/water the lining will be exposed to. For a long tunnel, such conditions might only apply locally, and the application of a sealing layer/sacrificial element of the lining should, therefore, also only be applied locally.
    - o Fibre-free layers on the inside of the tunnel, also known as regulating layers, can be required to cover a fibre reinforced lining to address maintenance/cleaning requirements or prepare for application of a membrane. Creating a smooth closed finish can require a high carbon mix even when compared with other sprayed linings. This should be considered when weighing up decisions on waterproofing strategy and secondary lining.
- One tunnel or two?
    - o This is a decision that will often be decided based on other discipline requirements including operations and maintenance. If both solutions work, it might come down to the tunnel designer to identify the better solution. The circumference of one large tunnel might be less than the combined circumference of two smaller tunnels. However, if you include any temporary supporting walls (see Chapter 6) needed during construction, the

carbon balance might tip the other way. A primary lining with many joints and reinforcement might be able to contribute less to the long term support, requiring a thicker secondary lining. Any carbon footprint comparison needs to be sufficiently detailed to capture all stages of the lining.

15.3. Reducing sources of waste

While the designer might argue that some sources of waste are inherent, or within the control of the construction team rather than themselves, the specification can be an important tool in maximising efficiency.

The 4th edition of the BTS *Specification for Tunnelling* allows greater flexibility in material choices to meet performance requirements than previous editions. This allows materials teams to source the most optimal mix and make the most of emerging technologies to remove/lower the CEM 1 content. There are other ways that the designer can have an influence:

- Mix open time. Allowing mixes prepared for one tunnel face to be kept open until they can be used for another. This will avoid mixes being dumped after each excavation and potentially avoid cleaning of the lines, pump and robot repeatedly.
- Face sealing. As with other sealing layers, the question should be asked as to whether this is required in all scenarios, or can its application be optimised. If no one is entering the tunnel face does a minimum thickness need to be stated, or can it be applied or not applied based on the site observations of the construction and supervision teams, supported by the daily review process, which react to data and upcoming locations of higher risk?
- Tolerances. Lining tolerances are often added to each other to create a lining that is set out considerably larger than the theoretical minimum required. With material factors within the design and the capacity of linings to redistribute loads, are all these tolerances required, or can they be approached differently?

emerald PUBLISHING

ice
Publishing

British Tunnelling Society
ISBN 978-1-83608-693-2
https://doi.org/10.1108/978-1-83608-690-120251018

Chapter 16
Specifications, assurance and design outputs

16.1. Specifications
16.1.1 Responsibilities

Consideration should be given to which party in the contract is responsible for the preparation, review, checking and approval of materials and workmanship specifications. This should form part of the overall division of SCL design responsibilities defined at the inception of the project. The interface between temporary works and permanent works should be carefully defined and understood by all parties.

At the earliest possible stage, the client should define any requirements as part of the brief to their designer either directly, if the client is responsible for design, or by way of the contractor in the case of design and build contracts.

At the point of drafting the specification, the author should have a full appreciation of the following:

- The contract obligations in terms of assurance – for example, the project manager, supervisor and appropriate roles within the contractor's organisation, so it is clear who can accept or reject work.
- The role of the designer's site representatives (e.g. shift coverage, supervision and responsibility).
- Definition of the contractor's assurance organisation so that self-certification roles and responsibilities are clear.
- For self-certification, a clear division of responsibilities.

The designer should produce materials and workmanship specifications covering the full range of tunnelling activities appropriate to the project. A single specification is preferable but project criteria may dictate separate specifications.

There should be clarity and consistency in approach at project interfaces (between adjacent contracts and designers; between different tunnelling or construction methods; between asset protection/surface monitoring). Key or complex interfaces should be managed using a formal document such as an interface control document (ICD), ensuring that each party has given the interface due consideration and that appropriate information has been shared.

16.1.2 Scope and content

Current best practice for tunnel design is the adoption of the BTS *Specification for Tunnelling*. This document provides a basic framework for designers upon which the designer can build the specific project requirements. The specification can consist of the BTS *Specification for Tunnelling* along with a list of additions, amendments and deletions. However, it is recommended that the body of the BTS *Specification for Tunnelling* text is amended to provide a complete specification document.

The designer should undertake a systematic review of the specifications so that they remain relevant and achievable, using feedback from previous projects.

Consideration should be given to the following:

- The key hold points within the tunnelling cycle so that setting out, profile/thickness control and material performance can be monitored through relevant inspection test plans (ITPs).
- The evaluation of early age strength in sprayed concrete and any limitations on excavation progress and/or personnel entry into exclusion zones.
- The pre-construction testing requirements for innovative details and/or performance requirements.
- The number and location of production tests considering the relative design risks and production impacts.
- Performance requirements of the waterproofing and quality and testing criteria to be employed during construction.
- The classification and management of cracks or other defects and the associated non-conformance report (NCR) processes.

The specifications should detail any particular parameters relevant to the design. These parameters should be compatible with the design assumptions and be aligned with the testing requirements stated in the specification.

16.2. Assurance
There should be careful consideration and clear definition of the roles and responsibilities of the relevant parties in the execution of the SCL works.

16.2.1 Design
The role of the client, contractor and designer should be formally specified. The role of the independent checker should be as set out in BS 6164:2019, clause 6.4.8. The recommendations in clause 6.4.10 of BS 6164:2019 regarding peer review should be followed. All parties to this process should have a clearly defined role, scope, authority and reporting structure so that each party's input to the development, review and approval of the final SCL design is well understood by all parties. Such arrangements should consider a balanced approach to maximise the benefit of independent expert advice and buildability input while avoiding the danger of design by committee.

16.2.2 Independent checking
Design checking should be undertaken in line with the requirements of clause 6.4 of BS 6164:2019 and the *Code of Practice for Risk Management of Tunnel Works* (The International Tunnelling Insurance Group, 2023). All SCL tunnel lining should be subject to independent category 3 checking. This is because the design requires an understanding of complex soil/structure interaction and considerable geotechnical interpretation. The consequences of an error in the design could be significant. For checking purposes, temporary and permanent works should be treated similarly (BS 6164:2019: clause 6.4.1).

The Category 3 checker should have a formal appointment and scope of services. Consideration should be given as to whether the checker is employed by the client, contractor or designer, since this has a significant bearing on the day-to-day management of the checker and their ability to communicate with, and be responsive to queries from, both the contractor and designer.

Effective design management is essential, and it is recommended that an engagement plan should be prepared defining how these parties would interact and communicate taking account of the requirements of BS 6164:2019, clause 6.8. An integrated design and checking programme should be prepared, together with a list of deliverables for delivery by both parties. This programme should be aligned with the contractor's programme, with suitable provision for float, particularly for complex SCL structures or areas of the design where disagreement between designer and checker has the potential to arise and the dispute resolution protocol in clause 6.4.8.6 of BS 6164:2019 may be invoked.

In order to maintain the independence of the checker, the information required by clause 6.4.8.3 of BS 6164:2019 should be made available to the checker. The conduct of the check should be in accordance with clause 6.4.8 of BS 6164:2019. The checker should be empowered to challenge the wider aspects of the design (e.g. compliance with Construction (Design and Management) (CDM) Regulations, ensuring that the residual design risks are ALARP by way of the consideration of alternatives, ensuring that the SCL design and specifications are aligned with any new materials or construction techniques, and ensuring that the drawings are sufficiently complete and detailed).

The checker should be kept abreast of any changes to the designer's or contractor's programmes and any re-sequencing of the works or changes in priority so that the category 3 checking can be completed in sufficient time to support procurement and construction activities.

16.2.3 Construction

The role of the client, contractor and designer should be specified in respect of supervision and inspection, with consideration given to the degree of the contractor's quality assurance self-certification combined with witnessing and/or inspection by the designer. A careful balance needs to be achieved between having too much responsibility and power resting solely with the contractor's organisation, and having over-inspection and scrutiny by too many duplicating layers of inspection teams.

The designer should consider what degree of site presence is required to carry out any design validation, feedback and interpretation (particularly in respect of displacement monitoring), and their role in the daily review meetings and RESS process. The designer should be required to do the following:

- Provide rapid response and adequate support to construction activities.
- Respond to the contractor's proposed changes with regards to implications for design.
- Resolve non-conformances and defects.
- Review materials compliance.

The client and designer's site teams together with the contractor should be briefed on their roles and responsibilities, so that the interfacing personnel have a clear understanding of each other's functions, duties and authorities on the project. The designer should share significant elements of the design with the contractor through briefings. The contractor should include the designer in relevant workforce briefings and processes to allow a common understanding of both parties' challenges and objectives.

The *Code of Practice for Risk Management of Tunnel Works* (The International Tunnelling Insurance Group, 2023) requires the client to ensure there is sufficient 'Independent Construction Supervision' during tunnel construction. It must be carried out by 'independent individuals or organisations'. This needs to be carried out by persons with sufficient competence and experience in risk management and in the construction of sprayed concrete linings. This is usually carried out by an independent organisation, but has on some projects been carried out by a separate department from within the contractor's organisation or by the designer, as long as sufficient independence, competence and experience can be demonstrated to the client and the insurers.

16.3. Design outputs

The SCL designer should have a clear brief and a clear understanding of the required deliverables and deadlines. The overall design programme should include suitable provision for the contractor's input, constructability reviews, independent checking and approvals. The programme should be supported by a detailed deliverable schedule with a specific date ascribed to each item.

The programme should be updated on a regular basis to incorporate any changes, and to facilitate consistency of understanding and expectation by the project team.

16.3.1 Drawings

While the design development increasingly occurs in the 3D modelling space it is important that information intended for construction is presented on assured 2D design drawings. While some aspects of the design can be communicated with 3D models, this is largely not possible for construction details.

Consideration should be given to the degree of detail that will be delivered by the designer and, if appropriate, whether any further details need to be supplied by others. This should be clearly stated and understood from the outset.

Where possible, the design and drawings should be developed with the contractor so that effectiveness and suitability of the contractor's plant and processes are taken into account.

Drawings typically fall into the following categories:

- General arrangements.
- Geological sections.
- Setting out general arrangements.
- Instrumentation and monitoring (I&M) general arrangements and detailed I&M cross-sections to be read in conjunction with the I&M plan with trigger levels.
- Assumed SCL construction sequences, including application of secondary linings.
- Tolerances, which should be clearly and unambiguously stated.
- Reinforcement details and bar bending schedules (primary and secondary linings, as applicable), including detailed requirements for areas of plain or fibre reinforced sprayed concrete.
- Details (i.e. joint details, waterproofing details).
- Toolbox item drawings (e.g. spiling, pocket excavation, well-points, face dowels etc.).

Drawings should be annotated with specific Safety, Health and Environmental (SHE) boxes that highlight the specific residual risks associated with the SCL structure. SHE boxes should focus on the key and/or unusual risks, or on risks that might not be readily apparent to the reader of the drawing. The SHE boxes should be cross-referenced to the CDM risk registers.

Formal constructability reviews should be held with the contractor's team, as the SCL design evolves, so that the final design aligns with the project's needs and is buildable. The designer and contractor should confirm the following points:

- That the overall tunnelling sequence depicted in the SCL design drawings is in accordance with the contractor's requirements, including spaceproofing for plant, ventilation and so forth, and considers intermediate conditions where access/egress may be impaired.
- The contractor's construction methods and other interfacing temporary works are compatible with the works shown on the SCL design drawings.
- The contractor is satisfied that the excavation and support sequences, and other relevant details shown, allow the contractor to discharge responsibilities for support and stabilisation of the excavations.
- The contractor has gained an understanding of the design basis of the SCL design.
- The contractor's proposed construction methods and temporary measures (or toolbox items) are compatible with and not detrimental to the SCL design.

It is recommended that such reviews and signoff are formally recorded.

16.3.2 Other documents

In addition to the suite of materials and workmanship specifications, the designer should prepare and issue the following documents to the client and contractor:

- A record of the SCL design, such as a basis of SCL design report or an SCL design statement.
- ICDs, for effective interface management at critical areas.
- Geotechnical data and interpretive reports, including factual geotechnical reports, selection of geotechnical design and modelling parameters and key assumptions.
- I&M requirements including trigger levels.

16.3.3 Risk assessments

With the integrated nature of SCL design and construction, and the severity of the risks often associated with tunnelling in general, it is essential that adequate attention is given to the effective and clear communication of residual design risks to the site teams, to allow continuity between the design and construction functions. The designer should not assume that formal issue of a CDM risk register to a contractor's organisation automatically results in the workforce and supervisors understanding the design drivers and the key residual design risks. While the primary responsibility for briefing the workforce should rest with the contractor, the designer should consider a range of methods to promote effective communication and understanding, including the following:

- Comparative risk assessments, to justify and explain the designer's decision-making process in producing an SCL design that has reduced the residual design risks to ALARP.
- CDM designer's risk registers, to highlight the risks inherent in the SCL design and how these have been eliminated, mitigated or transferred to another party.

- SHE boxes on construction drawings, highlighting the key residual risks specific to that element of the works.
- Highlighting significant residual risks on construction drawings through the use of mandatory or recommended temporary works; alternatively, the risk may be described in detail, together with a possible or suggested solution for design or implementation by others.
- Continuity of the designer's team through the construction phase to maintain understanding of the design decisions and basis.
- Designer-led toolbox talks, so that the construction, engineering, supervisory teams and the workforce understand the key residual design risks and the actions required of them.

British Tunnelling Society
ISBN 978-1-83608-693-2
https://doi.org/10.1108/978-1-83608-690-120251022

References

17.1. Key books and publications

Austin S and Robins P (1998) *Sprayed Concrete Properties, Design and Application.* Whittles Publishing, Dunbeath, Caithness, Scotland, UK.

BTS website recommended reading list: https://www.britishtunnelling.com

Crossrail Learning Legacy website: https://learninglegacy.crossrail.co.uk

Crossrail Best Practice Guide, SCL Exclusion Zone Management (2016).

Crossrail Project: Infrastructure design and construction, vol. 1 (2015).

Crossrail Project: Infrastructure design and construction, vol. 2 (2015).

Crossrail Project: Infrastructure design and construction, vol. 3 (2016).

HS2 Learning Legacy website: https://learninglegacy.hs2.org.uk/

Jones B (2022) *Soft Ground Tunnel Design.* CRC Press, London, UK.

Thomas A (2020) *Sprayed Concrete Lined Tunnels.* Taylor and Francis, UK.

17.2. Other reference documents

AFTES (2001) *Recommendations on The Convergence-Confinement Method*, https://tunnel.ita-aites.org/media/k2/attachments/public/Convergence-confinement%20AFTES.pdf.

AASCE (American Society of Civil Engineers) (2007) *Geotechnical Baseline Reports for Construction: Suggested Guidelines. Task Committee on Geotechnical Baseline Reports.* ASCE, Reston, VA, USA.

BRE (Building Research Establishment) Construction Division (2005) *BRE Special Digest 1.* BRE Electronic Publications.

BSI (British Standards Institute) (2016) PAS 8810:2016 Tunnel design. Design of concrete segmental tunnel linings. Code of practice. BSI, London, UK.

CIRIA (Construction Industry Research and Information Association) (2018) *C766 Control of Cracking Caused by Restrained Deformation in Concrete.* CIRIA, London, UK.

CIRIA (2020) *C791 The Management of Advanced Numerical Modelling in Geotechnical Engineering: Good Practice.* CIRIA, London, UK.

CIRIA (2023) *C807 Geotechnical Baseline Reports: A Guide to Good Practice.* CIRIA, London, UK.

Concrete Society (2008) *TR31 Permeability Testing of Site Concrete.*

LU (2016) Guidance Document GO55 Civil Engineering – Deep Tube Tunnels and Shafts. In-house document.

The International Tunnelling Insurance Group (2023) *Code of Practice for Risk Management of Tunnel Works*, 3rd edn, https://www.imia.com/wp-content/uploads/2023/10/Tunneling-Code-of-practice_Patrick-Bravery.pdf.

17.3. Papers and publications

Ahuja V and Jones BD (2016) Non-destructive approach for shotcrete lining strength monitoring. *Proceedings of the World Tunnel Congress, San Francisco.* SME, Englewood, CO, USA.

Anagnostou G and Kovári K (1996a) Face stability conditions with earth-pressure-balanced shields. *Tunnelling & Underground Space Technology* **11(2)**: 165–173.

Anagnostou G and Kovári K (1996b) Face stability in slurry and EPB shield tunnelling. In *Geotechnical Aspects of Underground Construction in Soft Ground* (Mair RJ and Taylor RN (eds)). Balkema, Rotterdam, the Netherlands, pp. 453–458.

Anagnostou G (2012) The contribution of horizontal arching to tunnel face stability. *Geotechnik* **35(1)**: 34–44.

Attewell PB and Woodman JP (1982) Predicting the dynamics of ground settlement and its derivatives caused by tunnelling in soil. *Ground Engineering* **15(8)**: 13–22 & 36.

Barratt DA, O'Reilly and Temporal J (1994) long-term measurements of loads in overconsolidated clay. *Tunnelling '94. Papers Presented at the Seventh International Symposium 'Tunnelling '94'*, London, UK, pp. 469–481.

Batty, E, Bond N, Kentish E, Skarda A and Webber S (2016) *Comparison Between Sprayed and Cast In-Situ Concrete Secondary Linings at Bond Street and Farringdon Stations*, https://learninglegacy.crossrail.co.uk/documents/comparison-sprayed-cast-situ-concrete-secondary-linings-bond-street-farringdon-stations/.

Bengt B and Broms M (1967) Stability of clay at vertical openings. *Proceedings of the ASCE*. ASCE, Reston, VA, USA.

Benz T (2006) *Small-Strain Stiffness of Soils and Its Numerical Consequences*. PhD thesis, Universitat Stuttgart, Germany.

Broms BB and Bennermark H (1967) Stability of clay at vertical openings. *Journal of the Soil Mechanics and Foundation Division* **93(1)**: 71–94.

Carter JP and Liu MD (2005) Review of the structured cam clay model. soil constitutive models: Evaluation, selection, and calibration. In *ASCE, Geotechnical Special Publication* 128. ASCE, Reston, VA, USA, pp. 99–132.

Chang Y and Stille H (1993) Influence of early-age properties of shotcrete on tunnel construction sequences. In *Shotcrete for Underground Support VI* (Wood DF and Morgan DR (eds)). ASCE, Reston, VA, USA, pp. 110–117.

Curtis DJ (1976) A discussion note. *Géotechnique* 26: 231–237.

DAUB (Deutscher Ausschuss für unterirdisches Bauen) (2016) *Recommendations for Face Support Calculations for Shield Tunnelling in Soft Ground*. DAUB, Köln, Germany.

Davis EH, Gunn RJ, Mair RJ and Seneviratne HN (1980) The stability of shallow tunnels and underground openings in cohesive material. *Géotechnique* **30(4)**: 397–416.

Dimmock PS and Mair RJ (2007) Estimating volume loss for open-face tunnels in London Clay. *Proceedings of Institution of Civil Engineers – Geotechnical Engineering* 160: 13–22.

Duddeck H and Erdman J (1981) Structural design models for tunnels. *Tunnelling* 82: 83–91.

Duddeck H and Erdmann J (1985) On structural design models for tunnels in soft soils. *Underground Space* 9: 246–259.

Einstein HH and Schwartz W (1979) Simplified analysis for tunnel supports. *Journal of Geotechnical Engineers* **105(4)**: 499–518.

Eisenstein Z and Branco P (1991) Convergence–confinement method in shallow tunnels. *Tunnelling and Underground Space Technology* **6(3)**: 343–346.

England GL and Illston JM (1965) Methods of computing stress in concrete from a history of measured strain – part 2: the rate of flow method. *Civil Engineering and Public Works Review* **(May)**: 692–694.

Farrell R and Terry D (2015) Pipe-jacked canopy performance at Bond Street, London, UK. *Proceedings of the Institution of Civil Engineers – Geotechnical Engineering* **168(3)**: 189–200.

Faustin NE, Elshafie MZEB and Mair RJ (2018) Case studies of circular shaft construction in London. *Proceedings of the Institution of Civil Engineers – Geotechnical Engineering* **171(5)**: 391–404.

Gakis A and Salak P (2016) Efficient design of openings in SCL tunnels. *Tunnels and Tunnelling International*: 41–46.

Hashash YMA, Hook, JJ, Schmidt, B and Yao JI-C (2001) Seismic design and analysis of underground structures. *Tunnelling and Underground Space Technology* **16**: 247–293.

Hoek E and Brown ET (1980) *Underground Excavations in Rock.* Taylor & Francis, London, UK.

ITA (International Tunnelling and Underground Space Association) (2011) *Monitoring and Control in Tunnel Construction*, Report No. 009/Nov 2011, https://about.ita-aites.org/publications/wg-publications/65/monitoring-and-control-in-tunnel-construction.

Jardine R, Potts D, Fourie A and Burland J (1986) Studies of the influence on non-linear stress-strain characteristics in soil-structure interaction. *Géotechnique* **36(3)**: 377–396.

John M and Mattle B (2003) Shotcrete lining design: Factors of influence. *Proceedings RETC.* Society for Mining, Metallurgy, and Exploration Inc., Littleton, CO, USA, pp. 726–734.

Jones BD and Grand C (2024) Strains in sprayed concrete tunnel linings at Heathrow Terminal 4 and back-calculation of stress. *Proceedings of the Institution of Civil Engineers – Geotechnical Engineering*, 'ahead of print', 10.1680/jgeen.23.00213

Jones BD, Davies AG and Ahuja V (2017) Sprayed concrete strength monitoring using thermal imaging at Bond Street Station Upgrade. *Proceedings of the World Tunnel Congress – Surface Challenges – Underground Solutions, Bergen, Norway.* Norsk Forening for Fjellsprengningsteknikk (Norwegian Tunnelling Society), Oslo, Norway, pp. 245–254.

Jones BD, Grand C and Clayton CRI (2023) Stresses in sprayed concrete tunnel linings at Heathrow Terminal 4. *Proceedings of the Institution of Civil Engineers – Geotechnical Engineering* **176(6)**: 646–661, 10.1680/jgeen.22.00054

Jones BD, Li S and Ahuja V (2014) Early strength monitoring of shotcrete using thermal imaging. *Proceedings of the 7th International Symposium on Sprayed Concrete – Modern Use of Wet Mix Sprayed Concrete for Underground Support, Sandefjord, Norway* (Beck T, Woldmo O and Engen S (eds)). Tekna & Norsk Betongforening, Oslo, Norway, pp. 245–254.

Karakus M (2007) Appraising the methods accounting for 3D tunnelling effects in 2D plane strain FE analysis. *Tunnelling and Underground Space Technology* **22**: 47–56.

Kielbassa S and Duddeck H (1991) Stress-strain fields at the tunnelling face. Three-dimensional analysis for two-dimensional technical approach. *Rock Mechanics and Rock Engineering* 24: 115–132.

Kimura and Mair (1981) Centrifugal testing of model tunnels in soft clay. *Proceedings of the 10th International Conference on Soil Mechanics and Foundation Engineering, Stockholm*, vol. 1. International Society for Soil Mechanics and Geotechnical Engineering (ISSMGE), University of London, London, UK, pp. 319–322.

King M (2018) Protection against fire for the UK Crossrail tunnel structures. In *Crossrail Project: Infrastructure design and Construction*, Volume 5 (Williams RV and Black M (eds)). ICE Publishing, London, UK, pp. 281–292.

Kirsch A (2009) Experimental and numerical investigation of the face stability of shallow tunnels in sand. *Acta Geotechnica* **5**: 43–62, 10.1007/s11440-010-0110-7.

Kovacevic N, Edmonds HE, Mair RJ, Higgins KG and Potts DM (1996) Numerical modelling of the NATM and compensation grouting at Redcross Way. *Proceedings of the International Symposium Geotechnical Aspects of Underground Construction in Soft Ground*, City University, London, UK (Mair RJ and Taylor RN (eds)). Balkema, Rotterdam, the Netherlands, pp. 553–559.

Kummerer C (2003) *Numerical Modelling of Displacement Grouting and Application to Case Histories*. Institute of Rock Mechanics and Tunnelling and the Institute of Soil Mechanics, Foundation Engineering and Computational Geotechnics, University of Technology Graz, Graz, Austria.

Leca E and Dormieux L (1990) Upper and lower bound solutions for the face stability of shallow circular tunnels in frictional material. *Géotechnique* **40(4)**: 581–606.

Lee SW, Bolton RJ, Hagivara T, Soga K and Dasari GR (2001) Centrifuge modelling of injection near tunnel lining. *IJPMG-International Journal of Physical Modelling in Geotechnics* **1**: 9–24.

Macht J, Lackner R, Hellmich C and Mang HA (2003) Quantification of stress states in shotcrete shells. In *Numerical Simulation in Tunnelling* (Beer G (ed.)). Springer-Verlag, Vienna, Austria, pp. 225–248.

Macklin SR (1999) The prediction of volume loss due to tunnelling in overconsolidated clay based on heading geometry and stability number. *Ground Engineering* **4**: 30–33.

Mair RJ (1994) Report on Session 4: displacement. *Proceedings of Conference on Grouting*. ICE, London, UK, pp. 375–384.

Mair RJ (2008) Tunnelling and geotechnics: new horizons. *Géotechnique* **58(9)**: 695–736.

Mair RJ and Taylor RN (1997) Bored tunnelling in the urban environment. State-of-the-art Report and Theme Lecture. *The 14th International Conference on Soil Mechanics and Foundation Engineering*, vol. 4. International Society for Soil Mechanics and Geotechnical Engineering (ISSMGE), University of London, London, UK, pp. 2353–2385.

Mair RJ, Gunn MJ, O'Reilly MP (1981) Ground movements around shallow tunnels in soft clay. *Proceedings of the 10th International Conference on Soil Mechanics and Foundation Engineering, Stockholm*, vol. 1, pp. 323–328.

Mair RJ, Taylor RN and Bracegirdle A (1993) Sub-surface settlement profiles above tunnels in clays. *Géotechnique* **43(2)**: 315–320.

Mair RJ, Taylor RN and Burland JB (1996) Prediction of ground movements and assessment of risk of building damage due to bored tunnelling. In *Geotechnical Aspects of Underground Construction in Soft Ground* (Mair RJ and Tayor RN (eds)). Balkema, Rotterdam, the Netherlands.

Vu MN, Broere W (2018) Structural design model for tunnels in soft soils: From construction stages to the long-term. *Tunnelling and Underground Space Technology* **78 (2018)**: 16–26.

Vu MN, Broere W, Bosch JW (2017) Structural Analaysis for Shawllow Tunnels in soft soils. *ASCE Int. J Goemech*, **17(8)**.

Moh ZC, Hwang RN, Fan CB and Chang JL (1997) Jacking up buildings by grouting. *Proceedings of 14th International Conference on Soil Mechanics and Foundation Engineering, Hamburg, Germany*. International Society for Soil Mechanics and Geotechnical Engineering (ISSMGE), pp. 1633–1636.

Möller S (2006) *Tunnel Induced Settlements and Structural Forces in Linings*. PhD thesis, Universität Stuttgart, Germany.

Möller SC and Vermeer PA (2008) On numerical simulation of tunnel installation. *Tunnelling and Underground Space Technology* **23**: 461–475.

Muir Wood AM (1975) The circular tunnel in elastic ground. *Géotechnique* **26**: 115–127.

Muir Wood D (1990) Anisotropic elasticity and yielding of a natural plastic clay. *International Journal of Plasticity* **6(4)**: 377–388.

Nasekhian A (2011) *Application of Non-probabilistic and Probabilistic Concepts in Finite Element Analysis of Tunnelling*. Dissertation, Graz University of Technology, Graz, Austria.

New BM (2017) *Settlements due to Shaft Construction*. Technical note, Harding Memorial Lecture.

New BM and Bowers K (1994) Ground movement model validation at the Heathrow Express trial tunnel. *Proceedings of the 7th International Symposium of the IMM and BTS*. Chapman and Hall, London, UK, pp. 310–329.

New BM and O Reilly MP (1991) Tunnelling induced ground movements: predicting their magnitude and effect. *4th International Conference on Ground Movements and Structures.*

O'Reilly MP and New BM (1982) Settlements above tunnels in the UK – their magnitude and prediction. In *Tunnelling '82* (Jones MJ (ed.)). Institution of Mining & Metallurgy (IMM), London, UK, pp. 173–181.

Panet M and Guenot A (1982) Analysis of convergence behind face of a tunnel. *Proceedings of Tunnelling '82.* IMM, London, UK, pp. 197–203.

Perazzelli P and Anagnostou G (2017) Analysis method and design charts for bolt reinforcement of the tunnel face in purely cohesive soils. *Journal of Geotechnical and Geoenvironmental Engineering* **143(9)**: 04017046.

Pferdekämper T and Anagnostou G (2022) Undrained trapdoor and tunnel face stability revisited. *Géotechnique Letters* **12**: 244–250.

Potts DM and Zdravkovic L (2001) *Finite Element Analysis in Geotechnical Engineering Application*, vol. 2. Thomas Telford, London, UK.

Pöttler R (1990) Time-dependent rock-shotcrete interaction, a numerical shortcut. *Computers and Geotechnics* **9**: 149–169.

Rokahr RB, Stärk A and Zachow R (2002) On the art of interpreting measurement results. *Felsbau* **20(2)**: 16–21.

Rokahr RB and Zachow R (1997) Ein neues Verfahren zur täglichen Kontrolle der Auslastung einer Spritzbetonschale. *Felsbau* **15(6)**: 430–434.

Rokahr RB and Zachow R (1999) *Betonspannungsermittlung in der Spritzbetonaußenschale mit dem Programm STRESS am Beispiel Eggetunnel*. Internal report, University of Hannover, Germany.

Schädlich B and Schweiger HF (2013) Influence of anisotropic small strain stiffness on the deformation behaviour of geotechnical structures. *International Journal of Geomechanics* **13(6)**: 861–868.

Schädlich B and Schweiger HF (2014) A new constitutive model for shotcrete. In *Numerical Methods in Geotechnical Engineering* (Group TF (ed.)). CRC Press, Boca Raton, FL, USA, pp. 103–108.

Schanz T, Vermeer P and Bonnier PG (1999) The hardening soil model: formulation and verification. *Beyond 2000 in Computational Geotechnics*. Taylor and Francis, London, UK, pp. 281–296.

Schweiger HF, Schuller H and Pöttler R (1997) Some remarks on 2-D-models for numerical simulations of underground constructions with complex cross sections. *Proceedings of 9th International Conference on Computer Methods and Advances in Geomechanics*, pp. 1303–1308.

St John AM, McGirr D, Weinmar W and Fornelli D (2016) *Design and Construction of Fisher Street Crossover Cavern on Crossrail Contract C300/C410*, https://learninglegacy.crossrail.co.uk/documents/design-construction-fisher-street-crossover-cavern-crossrail-contract-c300c410-london/.

St John AM, Potts VJ, Perkins OF and Balogh ZS (2015) Use of TBM pilots for large diameter SCL caverns. *Crossrail C300/C410, Proceedings of World Tunnel Congress, Dubrovnik, Croatia*. HUBITG (Croatian Society for Concrete Engineering and Construction Technology), Dubrovnik, Croatia.

Stärk A, Rokahr RB and Zachow R (2001) A new method for a daily monitoring of the stress intensity of a sprayed concrete lining. *Progress in Tunnelling after 2000, Proceedings of the World Tunnel Congress, Milan, Italy*, vol. 1 (Teuscher P and Colombo A (eds)). Patron Editore, Bologna, Italy, pp. 699–705.

Swoboda G (1979) Finite element analysis of the New Austrian Tunnelling Method (NATM). *Proceedings 3rd International Conference on Numerical Methods in Geomechanics, Aachen*, vol. 2. A. A. Balkema, Rotterdam, the Netherlands, pp. 581–586.

Taborda DMG, Pedro AMG, Xia H and Hardy S (2024) A methodology for improved predictions of surface ground movements around shafts. *Proceedings of the Institution of Civil Engineers – Geotechnical Engineering*, 10.1680/jgeen.24.00258.

Tucker NJ, Scevity K, Commins J and Linde E (2014) *Design and Construction of Crossrail Stepney Green Sprayed Concrete Lined Caverns*, https://learninglegacy.crossrail.co.uk/documents/design-and-construction-of-crossrail-stepney-green-sprayed-concrete-lined-caverns/.

Weiher H, Jones B and Runtemund K (2017) Early strength of shotcrete for tunnel advances – new monitoring approach using thermal imaging. *fib Symposium 2017 – High Tech Concrete – Where Technology and Engineering Meet*, Maastricht, the Netherlands. Springer.

Wongsaroj J, Soga K and Mair RJ (2007) Modelling of long-term ground response to tunnelling under St James's Park, London. *Géotechnique* **57(1):** 75–90.

Wongsaroj J, Soga K and Mair RJ (2013) Tunnelling-induced consolidation settlements in London Clay. *Géotechnique* **63(13):** 1103–1115.

emerald PUBLISHING ice

British Tunnelling Society
ISBN 978-1-83608-693-2
https://doi.org/10.1108/978-1-83608-690-120251023

Index

www.ingramcontent.com/pod-product-compliance
Lightning Source LLC
Chambersburg PA
CBHW081109220326
41598CB00038B/7287